Acronyms

EV	Enterprise Value
GE	General Electric
GHG	Greenhouse Gas
GNSS	Global Navigation Satellite System
GPS	Global Positioning System
ICAO	International Civil Aviation Organization
IP	Intellectual Property
IPO	Initial Public Offering
IT	Information Technology
ITAR	International Traffic in Arms Regulations
JDAM	Joint Direct Attack Munition
JV	Joint Venture
M&A	Mergers & Acquisitions
MOU	Memorandum of Understanding
MRO	Maintenance Repair and Overhaul
NASA	National Aeronautics and Space Administration
NATO	North Atlantic Treaty Organization
NDRC	National Development and Reform Commission
OEM	Original Equipment Manufacturer
PE	Private Equity
PNT	Positioning Navigation and Timing
PPP	Public Private Partnership
PRS	Public Regulated Service
R&D	Research & Development
ROIC	Return on Invested Capital
SME	Small and Medium-Sized Enterprise
UK	United Kingdom
US	United States
UTC	United Technologies Corporation
VC	Venture Capital
VTOL	Vertical Take-Off and Landing

ACRONYMS

A&D	Aerospace & Defense
ATAG	Air Transport Action Group
BAE	BAE Systems
C4ISR	Command, Control, Communications, Computers Intelligence, Surveillance and Reconnaissance
CDQM	Commercially Developed Military Qualified
CEO	Chief Executive Officer
CFO	Chief Financial Officer
CFRP	Carbon Fiber Reinforced Polymer
CMOS	Complementary Metal-Oxide Semiconductor
COO	Chief Operating Officer
COTS	Commercial-off-the-Shelf
CTO	Chief Technology Officer
CVC	Corporate Venture Capital
DARPA	Defense Advanced Research Projects Agency
DJI	Dow Jones Industrial Average
EBIT	Earnings Before Interest Tax
EBITDA	Earnings Before Interest Tax Depreciation and Amortization
EPA	Environmental Protection Agency
EPS	Earnings per Share
EU	European Union

ABOUT THE AUTHOR

ANTOINE GÉLAIN is a corporate strategy advisor, expert in the Aerospace & Defense industry, having worked for thirty years in it, first as an international sales manager in Asia-Pacific for MATRA and AEROSPATIALE (now Airbus Group), then as a strategy consultant advising corporate clients and private equity investors active in the sector.

He started his consulting career with BOOZ ALLEN HAMILTON, before launching his own consultancy – PARAGON EUROPEAN PARTNERS. He also runs the A&D practice of CANDESIC, a London-based boutique strategy consulting firm.

Antoine holds a master's degree from ESSEC School of Management in France and an M.B.A. from HARVARD BUSINESS SCHOOL. He is a Fellow of the Royal Aeronautical Society (FRAeS) and a featured columnist for Aviation Week & Space Technology, where he writes on business strategy and innovation. Born and raised in France, married to an American, he has been living in London, UK for more than two decades.

He can be contacted at www.thebusinessofaerospace.com.

The Business of Aerospace

*Industry Dynamics, Corporate Strategies,
Innovation Models and...the Big(ger) Picture*

The Business of Aerospace

Industry Dynamics, Corporate Strategies, Innovation Models and...the Big(ger) Picture

Antoine Gélain

Paul Park, Editor-in-Chief
University of Texas at Arlington
Arlington, Texas

Published by
American Institute of Aeronautics and Astronautics, Inc.
12700 Sunrise Valley Drive, Reston, VA 20191-5807

American Institute of Aeronautics and Astronautics, Inc., Reston, Virginia

1 2 3 4 5

Library of Congress Cataloging-in-Publication Data

Names: Gélain, Antoine, author.
Title: The business of aerospace : industry dynamics, corporate strategies, innovation models and...the big(ger) picture / Antoine Gélain.
Description: Reston, VA : American Institute of Aeronautics and Astronautics, Inc., [2021] | Series: Library of flight | Includes index.
Identifiers: LCCN 2021022220 (print) | LCCN 2021022221 (ebook) | ISBN 9781624106149 (hardcover) | ISBN 9781624106156 (ebook)
Subjects: LCSH: Aerospace industries–Management.
Classification: LCC HD9711.5.A2 G39 2021 (print) | LCC HD9711.5.A2 (ebook) | DDC 629.1068–dc23
LC record available at https://lccn.loc.gov/2021022220
LC ebook record available at https://lccn.loc.gov/2021022221

ISBN: 978-1-62410-614-9

Copyright © 2021 by Antoine Gélain. Published by the American Institute of Aeronautics and Astronautics, Inc., with permission. No part of this publication may be reproduced, distributed, or transmitted, in any form or by any means, or stored in a database or retrieval system, without the prior written permission of the publisher.

Data and information appearing in this book are for informational purposes only. AIAA is not responsible for any injury or damage resulting from use or reliance, nor does AIAA warrant that use or reliance will be free from privately owned rights.

In memory of my parents,
Jean-Marie (1929–2019) and Geneviève (1931–2019)

Contents

Foreword xi

Preface xiii

Introduction xv

Part One   Industry Dynamics 1

Chapter 1   Industry Structure and Competition 3

Chapter Introduction 3
Private Equity's Promise 7
The Big White Space of the European A&D Industry 11
The New Predators 15
A and D Divergence 18
Digital Aerospace 21
Aerostructures: The Next Battleground 24
A Duopoly on the Edge 27

Chapter 2   Supply Chain Dynamics 33

Chapter Introduction 33
Time to Bridge the Industry – Finance Gap 35
Success Hard to Tell from Panic 37
Chain Drive 40
Revisiting the Customer-Supplier Relationship Model 43

Chapter 3   (Geo)Politics 47

Chapter Introduction 47
China's Aerospace Dream Edges Closer to Reality 48

A Proud Past but Uncertain Future (UK Aerospace). 50
Merkel's Mark 52
British A&D in Shambles 54
Galileo's High Stakes. 57

PART TWO   CORPORATE STRATEGIES 61

CHAPTER 4   BUSINESS FUNDAMENTALS 63

Chapter Introduction 63
Culture is Destiny 65
Clusters of Vitality 68
Closing the Deal 71
Managing the Political Risk. 74

CHAPTER 5   GROWTH STRATEGIES. 79

Chapter Introduction 79
Shrinking to Grow 81
Airbus Shows the Way. 84
Merger Logic. 87
Hope and Despair. 90

CHAPTER 6   CORPORATE STORIES 93

Chapter Introduction 93
Out of Steam 95
Dassault's Secret Sauce. 98
How L3 Lost its Magic. 101
Airbus at 50. 106
Doomed Strategy 110
Fallen Icon. 113

PART THREE   INNOVATION MODELS AND…THE BIG(GER) PICTURE. 117

CHAPTER 7   STRATEGIC INNOVATION 119

Chapter Introduction 119
Disruptive Innovations. 120
Creative Destruction 124
Comfort Zone. 127
Disruptive Imperative 132
Airbus's Innovation Gamble. 135

CHAPTER 8   NEW SPACE 139

Chapter Introduction 139
Separating Hype from Reality 141
When Musk Meets Disney 143
Game Over 146
Nanosatellites: Time to Deliver 150
Colonizing Mars 154

CHAPTER 9   THE BIG(GER) PICTURE 157

Chapter Introduction 157
The Lost Art of Air Travel 159
Dirty Secret 161
Smaller, Slower, Closer 164
What is the Purpose of Space Exploration? 167
The Age of Resilience 170

CONCLUSION 173

A Post-COVID-19 Agenda for Industry Revival 174

INDEX 181

ACKNOWLEDGEMENTS 187

ABOUT THE AUTHOR 189

ACRONYMS 191

FOREWORD

For eight years, Antoine Gélain has provided Aviation Week's readers with thoughtful insights about the business of aerospace and defense. His concise 750-word columns look at our industry and its technological achievements through the eyes of a pragmatic businessman. Written from London, the columns often focus on European aerospace. But they all have a global backdrop, infused with Antoine's understanding of the interconnected nature of the industry and the borderless strategies that are crucial to innovation and reinvention.

Viewing his columns together in an organized ordering – as they are in this book – makes them even more powerful. Antoine may hail from Europe, but he is anything but parochial. He calls it like he sees it, as a number of irritated CEOs or government officials will confirm. He has written often about how some European aerospace companies lag their North American peers in value creation and offers strategies they should be adopting to catch up. His columns are reminiscent of the blunt observations of the late Pierre Sparaco, Aviation Week's legendary Paris bureau chief.

No company is too big to escape Antoine's scrutiny. But he is always fair, backing up his conclusions with reasoned points. And when praise is deserved, he gives it. In recent years, much of his praise has been focused on space, where a new generation of companies have disrupted what had become a staid industry with radical new visions and business models, from nanosats to dreams of colonizing Mars.

As the pace of disruptive innovation accelerates and sustainable aviation moves from "nice-to-do" to "must-do," the business decisions and technology investments companies and governments make today are more important than ever. Those who remain complacent will soon be obsolete. This book provides much to think about as our industry heads toward its next Golden Age.

Joe Anselmo
Editorial Director, Aviation Week Network
Editor-in-Chief, Aviation Week & Space Technology

PREFACE

Apart from my grandfather spending some time as an observer on board a Maurice Farmand MF.11 Shorthorn reconnaissance airplane during the First World War, nothing in my family predisposed me to get involved with aviation. And unlike many people who end up working in the aerospace industry, I did not grow up dreaming of becoming an astronaut or a fighter pilot nor did I study engineering.

But, after spending a summer as a blue-collar worker in Japan and a year as a financial analyst in South Korea, I fell in love with the region and my first goal back in Paris was to get a job that would take me back to Asia. That is how I ended up joining an Aerospace & Defense (A&D) company as a business development and export sales manager in charge of Asian markets. This company was Aerospatiale, the leading French aerospace group at the time, which first merged with Matra to become Matra Aerospatiale, then with DaimlerChrysler Aerospace (DASA) to become EADS, now Airbus Group. And so, these years selling aerospace products (tactical missiles such as the famous Exocet if it must be said) in Asia started what turned out to be a 30-year career, first working in, then consulting with the aerospace industry.

Indeed, after selling missiles for a while and honing my understanding of business and management at Harvard Business School, I became a strategy consultant specialized in the A&D industry, advising large and small companies as well as financial investors active in the sector. At some point, I started being quoted in various aerospace publications, including Aviation Week & Space Technology magazine, for which one-off quotes and articles became regular columns, on a wide range of topics, but always about the business side of aerospace, and in most cases from a strategy or innovation angle.

This book is a compilation of such columns written over the last decade or so. They are organized by themes and complemented occasionally by short commentaries introducing underlying business concepts or additional information.

These columns are personal viewpoints, written at specific points in time on topics pertaining to the aerospace business (or – more broadly defined – to the "Aerospace & Defense" business). As such they are a highly subjective (and European) take on the industry. However, with hindsight, I believe they collectively cover a broad enough range of issues to provide a comprehensive – if not exhaustive – view of the key themes relevant to the business of aerospace today.

My main purpose in completing this book is to help aerospace engineering students and young professionals better understand the business side of the industry while familiarizing themselves with generic business strategy and innovation concepts. To my knowledge, little has been written about the business of aerospace. Most books about the industry are either technical books or historical accounts of iconic companies or products. Yet, aerospace is not only about engineering and technology. It is a fully-fledged business with its own economics, competitive dynamics, key success factors and managerial challenges. And it is a complex business, not only because every airplane or rocket is made up of millions of parts produced by thousands of companies all over the world, but also because it is considered a "strategic" industry by many nations and therefore subject to a political powerplay that often interferes with rational business and managerial decision-making.

Even though some of the columns were published a while back, I did not update them, as they are meant to reflect a specific view taken or analysis made whenever they were written. A such, they are like short case studies, reflecting how business decisions are made: in the present moment, trying to make sense of the past and anticipate the future. Whether they turn out to be right or wrong, what matters most is to understand why and how they were made, so that lessons can be learned.

INTRODUCTION

As I am putting this book together, the world is still coming to terms with the COVID-19 pandemic. Air travel is down 40% compared to the previous year, many airlines and aerospace suppliers are on the brink of bankruptcy and even large primes like Boeing and Rolls-Royce are struggling to stay afloat.

Nobody knows yet how all this will play out over the next months and years. Of course, many of us in the aerospace community are trying to estimate how big the overall damage will be and how long it will take for the industry to get back to some level of normalcy. But this exercise is somewhat futile because in extreme, "out-of-boundary" situations like the one we are living through, standard forecasting models and usual economic models do not work anymore. It is like doing division to derive a ratio. When the denominator gets (close) to zero, the result (infinity) becomes meaningless.

And so, we are left with two yardsticks: our own gut feeling and our ability to take some altitude and look at the situation from a distance to be able to see some underlying patterns or long-term trends that may emerge from the apparent chaos.

Looking at commercial aerospace this way, several things become clearer. First, one must recognize that its growth trajectory before the crisis was unsustainable, with a lot of self-deception and structural weaknesses embedded in the system. When the industry starts moving from survival to recovery mode, it will need to address those weaknesses and "reset" expectations about the future. That could be the "**great corrector**" effect, applied to aerospace.

A big correction could happen on the demand side. While most analyses point to a recovery timeframe from 18 months to three years (back to 2019 levels and then back to the historical growth rate), several variables could further dramatically impact the future air traffic growth trajectory. Those variables include:

- Increased environmental pressure, e.g., will governments impose sustainability covenants to airlines as a condition for their aid?
- Consumer behavior, e.g., will consumers travel as much and with the same level of tolerance for "poor" service as before?
- Airline business models, e.g., will low-cost airlines continue to push the sector towards commoditization, or will there be a revival of a highly differentiated and segmented market?

Such variables could collectively reverse the shape of the air traffic growth curve from its historical convex shape to a concave one, as the long-term growth rate slows down significantly. In one such scenario, by 2035, Revenue Passenger – Kilometers (RPK) could be one third lower than the 15 trillion expected before the crisis. This would imply a long-term annual growth rate between 1995 and 2035 of around 3.8%, which, all things considered, would not be that bad (and would improve the odds of the industry achieving its sustainability targets), but still much lower than the pre-COVID 4.9% forecast.

Another correction could and should happen on the supply side. The aerospace supply chain is incredibly complex and inefficient, with parts and sub-assemblies going back and forth between thousands of players scattered all over the world, partly because of history (in Europe, Airbus was first a political construction, and then became an industrial puzzle) and partly because of the bad habit of tying up airplane sales with industrial offsets. With a market growth possibly slowing down by as much as 50% in the medium-term, the time has come for all large OEMs to reset their supply chain strategy and footprint.

What is needed is simpler, more compact, and more modular supply chains with a much deeper level of cooperation between customers and suppliers. In that context, governments, which are back in the driver's seat as they deploy their stimulus toolkit, will play a key role in setting the rules of the game, hopefully with a proper industrial strategy in mind. In fact, the way each government will intervene could well be the main driver of competitive advantage or disadvantage for their country's industry going forward.

Apart from being a "great corrector", this crisis is likely be a "**great accelerator**" too, as it crystallizes forces that were already at play before it started.

The marginalization of traditional (predominantly hardware) aerospace products in the digital economy is one of those forces. Value will migrate at an even faster pace towards software and service industries, and digital communications and AI-driven computing will replace transportation as the key enablers of tomorrow's economic activity. This may seem far-fetched, but when people and companies will be able to do 90% of their job remotely and digitally, from controlling 4.0 factory floors to managing a fleet of mining robots, transportation systems such as large commercial airplanes will become a footnote in their ecosystem.

Above all, the crisis will likely accelerate the rise of Asia and of China in particular, with its corollary: the decline of the West, where 90% of the aircraft manufacturing industry is still located. From a 'competitive dynamics' standpoint, it probably means difficult times ahead for the Airbus-Boeing duopoly, which shortcomings are laid bare today: Despite their decades-old arch-dominance, blinded by their head-to-head rivalry, they just took too many long-term risks for short-term wins. The recent Boeing's 737MAX $5 billion fiasco and Airbus' $4 billion corruption inquiry settlement illustrate their reckless behavior.

Meanwhile in Asia, after playing Airbus and Boeing against each other and benefiting from the "laissez-faire" of Western nations, their Chinese competitor finds itself in the strongest position ever. It now has all the liquidity, the domestic market scale, the political efficiency, and the industrial capacity necessary to accelerate its quest to become an aerospace powerhouse. The West on the other hand, will likely come out of the crisis financially bled dry, industrially bruised, and politically in shambles.

In the space and defense sectors, the dynamics should be somewhat different as institutional customers are expected to play a stabilizing role, thus mitigating the impact of the crisis. However, one should expect both some corrections (e.g., few of the many planned satellite constellations should materialize) and some acceleration of pre-existing trends as well (e.g., "New Space" players displacing "Old Space" ones).

A looming question also overclouds the defense and space sectors: Will institutional customers (defense ministries and space agencies) have the financial means and political leeway to keep funding their defense procurement and space programs at pre-crisis levels? This may prove difficult when so much public money and effort will be needed to rescue other economic sectors that may be given higher priorities in the wake of the COVID-19 pandemic, such as healthcare, consumer finance and education.

In any case, one can safely predict that the business of aerospace will become more challenging over the next ten years than it has been in the last ten. And this makes the contents of this book even more timely and relevant. Because it focuses on the fundamentals of the aerospace business - industry dynamics, corporate strategies, and innovation models – the stories it relates and the lessons and insights drawn from them are timely references to point a way forward, towards a revived success and a more resilient future. Indeed, as I wrote, back in 2019, in what is the final column of this book, I believe the aerospace industry is entering a new age, an age *"defined not so much by technological prowess, corporate wealth accumulation or relentless growth, as has been the case for the last 30 years, but rather by restraint, humility and resilience"*. It was foreseeable before the COVID-19 crisis and it is now a certainty.

And so, the final column of the book is even more relevant now and the final quote, from French philosopher Edgar Morin, even more powerful. For all of us, it should be food for thought…and action.

* *
*

The book is divided in three parts and nine chapters:

Part One: Industry Dynamics

1. **Industry Structure and Competition**: The chapter's first three columns were written roughly six years apart from each other (2007, 2013, 2020) on a broadly similar topic: the fundamental differences in the industry structure between the US and continental Europe and how it impacts the competitiveness and the dynamics of each region's A&D industry. The rest of the chapter addresses more diverse topics relevant to industry dynamics and competition: the increasing divergence between the aerospace and defense sectors; how the digital economy is seeping through every corner of the A&D industry, putting the traditional players at risk; why the aerostructures sector is likely to become the next commercial aerospace battleground; and, finally, the state of the Airbus-Boeing duopoly.

2. **Supply Chain Dynamics**: The aerospace industry relies on an overly complex supply chain, with thousands of small and medium-sized subcontractors involved in making millions of parts on all continents. This

INTRODUCTION						xix

chapter covers the key challenges they face, be it in terms of financial resources, growth, operational performance, or customer relationships. It looks at such challenges both at individual company level (supplier and customer) and at the whole supply chain or "system" level.

3. **(Geo)Politics**: The A&D industry is highly political and strategic in many countries, starting with those with the most at stake, namely the US, Western Europe, China and Russia. This chapter illustrates this geopolitical dimension by putting in contrast the ambitions of a new aerospace power – China – with the struggles of an old one – The United Kingdom. It also contrasts the pragmatism of Germany's export policy with the failings of France's political leadership. Overall, it shows how political decisions impact each country's industry success and, specifically for the UK, how the ongoing exit from the European Union (EU) has created a new challenge for both the British industry and the EU as political dreams clash with economic reality.

Part Two: Corporate Strategies

4. **Business Fundamentals**: This chapter covers four key strategic themes that constitute fundamental drivers of competitiveness in general and specifically in the A&D industry: corporate culture, geographic clusters, commercial craftsmanship (or the "art and craft" of selling aerospace products), and the ability to manage the political risk inherent to the aerospace business.

5. **Growth Strategies**: This chapter discusses and illustrates two key corporate strategy dimensions: business portfolio management and Mergers & Acquisitions (M&A) and looks at their underlying drivers and strategic value through the lenses of specific company examples (Thales, General Dynamics, Airbus Group) and M&A cases (Harris-Exelis and Safran-Zodiac).

6. **Corporate Stories**: This chapter is a collection of case studies on the strategic successes and failures of a few iconic A&D companies: BAE Systems, Dassault Aviation, L-3 Communications, Airbus, Bombardier, and Rolls-Royce. In each case, key strategic challenges or decisions are highlighted, analyzed, and commented.

Part Three: Innovation Models and…the Big(ger) Picture

7. **Strategic Innovation**: This chapter is about innovation in the A&D industry, whether from the point of view of Clay Christensen's disruptive

innovation theory – or from a more traditional angle (R&D funding models, "closed" innovation…). It also looks at what it would take to disrupt the commercial aerospace ecosystem currently monopolized by Boeing and Airbus. Finally, it gives an assessment of Airbus Group' recent attempt at revamping its global innovation strategy in a dramatic – potentially traumatic – way.

8. **New Space**: This chapter is a direct follow-on of the previous one, focusing on innovation dynamics in the space sector and more specifically what is called "New Space", to reflect the development of a new generation of players coming in with radically new visions, business models and value propositions. Over the last decade, the space sector has attracted technology disruptors, visionary entrepreneurs, and long-term investors alike, to the point where it has become the fulcrum of disruptive innovation in the aerospace industry.

9. **The Big(ger) Picture**: This final chapter is meant to step back from the purely business aspects of aerospace and addresses what I regard as more fundamental questions about the future of the industry. From the lost art of air travel to the challenge of sustainability, from the future of commercial aviation in the context of individual mobility systems to the deeper meaning of space exploration, these issues are bound to shape the industry in the decades to come and as such are worth reflecting upon.

Conclusion: A Post-COVID-19 Agenda for Industry Revival

The COVID-19 crisis has been shaking the industry's foundations and laying bare its weaknesses. It is therefore critical for all stakeholders to step back and think about the structural factors that will allow the industry and its constituents to spring back to success, this time in a more sustainable way. It entails avoiding the pitfalls of the past and setting a new agenda for a more resilient future. This concluding chapter lays out such an agenda for industry revival, drawn from all the stories related in this book.

Part One

INDUSTRY DYNAMICS

Chapter 1

INDUSTRY STRUCTURE AND COMPETITION

"One man's meat is another man's poison"

Lucretius, Roman poet and philosopher (c.99 – c.55 BC)

CHAPTER INTRODUCTION

The chapter's first three columns were written roughly six years apart from each other (2007, 2013, 2020) on a broadly similar topic: the fundamental differences in industry structure between the US and Europe and how it impacts the competitiveness and the dynamics of each region's A&D industry. Because I feel it is a particularly important theme in understanding industry dynamics, let me start by putting some context around it.

In 2007, the diagnostic was that European A&D companies' poor performance compared to their American peers was mainly due to the rigidity and disbalance of the European industry structure, compared to the fluidity and modularity of the American one.

In the US, the industry had been reconfiguring itself from a historical structure defined by a fragmented and platform-centric competitive landscape, a high level of vertical integration, proprietary product and system architectures, limited interface points with end customers, and a stable hierarchy between the various levels of the industry pyramid (*see left pyramid in chart below*). The new structure that was emerging then was characterized by less vertical integration, more modularity, more interface points with customers, and new players fighting to take positions in the middle of the pyramid, which became a "turbulence zone" (*see right pyramid in chart below*).

This reconfiguration process was driven by two major factors: large primes span off significant chunks of their business portfolios (mainly upstream activities such as components and minor modules and sub-systems)—more than $15 billion worth between 1995 and 2005 for the top five US primes alone; and private equity investors came in and played the role of catalyst for the emergence of a new breed of independent suppliers, positioned between primes and traditional subcontractors.

THE U.S. A&D INDUSTRY RECONFIGURATION PROCESS
(As of 2007)

Source: Author

These private equity-backed players implemented similar strategies, building "horizontal" portfolios of businesses with a consistent value chain positioning (at the level of components and small sub-systems), focusing on financial discipline and operational excellence, and promoting a fresh business approach of independent suppliers with a critical mass of R&D and marketing skills. By doing so, they contributed significantly to the restructuring and the competitiveness of the US A&D industry in the late 1990's and early 2000's.

In continental Europe, none of this was happening. The industry structure remained rigid, unbalanced and underperforming. Large companies, which were still perceived as being in the driver's seat for industry consolidation, were unable or unwilling to adopt the same type of dynamic business portfolio management approach, thus continued to rely on captive supply chains and a high level of vertical integration, which was becoming increasingly harmful to their performance. Any consolidation among the primes was essentially a zero-sum game.

That is why, in my first 2007 column, I was calling for two things to happen in Europe: First, large companies needed to invite qualified private equity investors to take over some of their non-core businesses. Second, private equity investors themselves needed to raise their interest in European A&D and start investing significantly and consistently to help new intermediate players emerge like in the US

In 2013, I revisited this theme in the wake of the failed EADS-BAE Systems merger, arguing again that what Europe needed to be more competitive was not bigger primes but more diverse, independent players in the middle of the pyramid, where there was still a "big white space" (*see chart in the column itself*). I reiterated my call for top European players to manage their portfolios more dynamically, making sure all their businesses gain from their ownership, spinning out those that do not and letting them become growth platforms themselves.

In 2020, I looked at how several US "horizontal" players dedicated to "components and subsystems" had been building huge business portfolios and creating a lot of value for themselves, their shareholders and – I assume – their customers. The interesting fact was that these companies, while following the value creation model initiated by private equity funds a decade earlier, were all publicly listed. While noticing that no similar companies existed in Europe, it made me realize that the traditional private equity model as practiced in Europe was probably too short-term and too short-sighted to play the role that I was calling for thirteen years ago.

Instead, either the reconfiguration of the European industry would be achieved by American players such as the ones described – with potential side effects on Europe's sovereignty and technological base, or it would not happen until large European groups articulate a proper "component and subsystem" strategy. The goal of such strategy should be to nurture this new breed of players that Europe -more than ever – desperately needs if it wants to remain competitive and independent in the long-term.

The chapter's last four columns address more diverse topics relevant to industry dynamics and competition: one is the increasing divergence between the aerospace and defense sectors of the industry, thus making me wonder whether it is still relevant to talk about the "Aerospace & Defense" industry.

Another one is about how the digital economy is seeping through every corner of the A&D industry, putting the traditional industry players at risk of becoming marginalized and their products – ultimately – commoditized.

The chapter's last but one column explains why the aerostructures sector is likely to become the next commercial aerospace battleground, thus triggering a complete reshuffling of the underlying supply chain and a huge challenge for all existing suppliers.

The final column takes stock of where the Boeing-Airbus duopoly stands in the wake of the COVID-19 crisis and what the future holds for both players, in particular in the context of the impending entry of a new competitor from China.

Private Equity's Promise

June 2007

MANY INSIGHTS CAN BE DRAWN from this year's Aviation Week's Top-Performing Companies (TPC) rankings but let me jump straight to a conclusion: As far as continental Europe is concerned, the time has come to shake things up and let a new generation of aerospace & defense (A&D) players emerge. For this to happen, Europe badly needs fresh ideas and money, and this can only come from private equity. What has already happened in the US in that respect should encourage those who want Europe to regain competitiveness and should reassure those who fear private equity greed.

Private equity players have been active in the US A&D industry for more than a decade, creating a new breed of companies and helping restructure complete sectors.

Private equity players have been active in the US A&D industry for more than a decade, creating a new breed of companies and helping restructure complete sectors. The Carlyle Group has to a large extent triggered the consolidation of the US aerostructures sector by acquiring multiple businesses, ultimately leading to the revival of Vought Aircraft Industries (see chart below). Onex is now following a similar path with Spirit Aerosystems. Veritas Capital has made multiple investments that have led to the establishment of such a successful defense electronics player as Integrated Defense Technologies, purchased by DRS Technologies in 2003. Odyssey Investment Partners played a similar catalyst role by aggregating multiple small aircraft component suppliers into what became publicly traded TransDigm. L-3 Communications itself, which best embodies the new generation of US A&D players, was the result of a private equity deal between Lockheed Martin, Lehman Brothers Private Equity, and the management. In these instances, private equity either played the role of catalyst, aggregator, or consolidator in the context of a desperately needed industry restructuring. Of course, this reconfiguration of the US landscape would not have been possible without large primes taking a fresh look at their business portfolios and spinning off significant chunks—more than $15 billion in the last 10 years for the top five US primes alone.

Most of these private equity-backed players have employed similar strategies, building up "horizontal" portfolios of businesses with a consistent value chain positioning, focusing on financial discipline and operational excellence, and promoting a fresh business approach of independent suppliers with a critical mass of R&D and marketing skills. They are thus capable of rapidly bringing to market innovative high-performance products and services.

THE CARLYLE GROUP'S CATALYST ROLE IN THE CONSOLIDATION OF THE U.S. AEROSTRUCTURES SECTOR

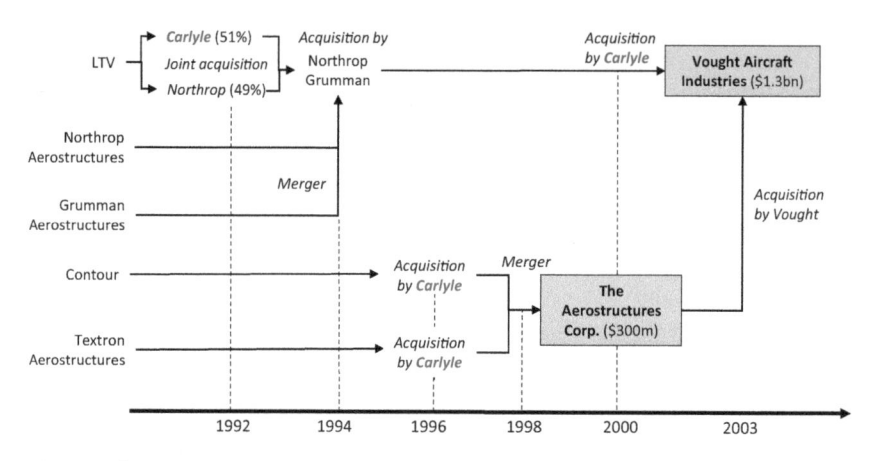

```
LTV ──┬─► Carlyle (51%) ──┐ Acquisition by        Acquisition
      │  Joint acquisition │ Northrop              by Carlyle      Vought Aircraft
      └─► Northrop (49%) ──┘ Grumman ──────────────────────────►   Industries ($1.3bn)

Northrop
Aerostructures ─────────────────┐
                                 │ Merger
Grumman                          │                                  Acquisition
Aerostructures ─────────────────┘                                   by Vought

Contour ─────────────────────► Acquisition   Merger
                                by Carlyle ──────┐    The
                                                  ├─► Aerostructures
Textron                        Acquisition       │    Corp. ($300m)
Aerostructures ──────────────► by Carlyle ──────┘

              1992      1994       1996      1998      2000       2003
```

Source: Author

In Europe, there is a critical lack of independent players with the proper mix of technological scope, industrial scale, and marketing clout.

In continental Europe, this has not happened. The value chain remains rigid, unbalanced and underperforming. Large companies, which are still perceived as being in the driver's seat for industry consolidation, are constrained by a mix of industrial legacy, political interference and strategic "myopia." Below them—aside from a few success stories such as Indra Sistemas or Ultra Electronics—there is a critical lack of independent players with the proper mix of technological scope, industrial scale and marketing clout. This is where the stakes are high for European industry and where private equity money is needed. If Europe does not find a way to ensure the emergence of intermediate players in such critical sectors as defense electronics or aircraft manufacturing, large groups will become less competitive and small suppliers will not survive the onslaught of global competition.

Two things must happen fast. First, large companies need to start welcoming qualified private equity investors to take over some of their non-core businesses through complete or partial spin-offs. Second, private equity funds need to raise their interest in European A&D and start investing significantly and consistently. The challenge is finding a way around a Catch 22 situation. Large companies will not start spinning off operations until they feel confident it will not endanger their supply base. But private equity funds aren't going to start investing large sums in the industry until they feel confident that the deal flow is sufficient to support ambitious build-up strategies that would enable them to create players capable of competing globally with the likes of L-3 DRS, Spirit and Vought.

Let's face it: recent consolidation among European national primes has essentially been a zero-sum game, and it is likely to remain so if they keep swapping assets among themselves. It is therefore necessary to apply new business models and bring in fresh money. And today, this means private equity. These investors should play a key role in the European A&D industry reconfiguration by partnering with large groups to "spin off" some of their businesses. They also can help smaller, under-funded companies reach full potential by migrating their business models and mindsets from one of a "slaved vertical sub-contractor" to one of an "independent horizontal supplier."

Ultimately, this should allow a new breed of globally competitive European player to fill in the industry white space, through either traditional scale-driven industrial consolidation or more innovative portfolio-like aggregation, thus contributing to the overall efficiency and sustainability of the European A&D supply base.

> **KEY CONCEPT**
>
> ### INTEGRATION VS. DIS-INTEGRATION CYCLES
>
> Most industries go through **cycles of** (vertical) **integration** and **dis-integration** (or 'de-verticalization'), depending on the predominant basis of competition. These cycles have a significant impact on the way value is created by and shared among the industry participants.
>
> In general, when the main basis of competition (i.e., what customers value the most) is the ability to offer a functionality (e.g., flying from A to B for a commercial aircraft, or destroying a target for a weapon system) with an acceptable level of reliability (such as 99% safety performance for a commercial aircraft, or 'single shot kill probability' for a weapon system), vertical integration is an advantage as long as customer needs have not been satisfied along this performance dimension.
>
> In this context, companies have a better chance of achieving the expected level of performance by controlling all the interfaces within the product architecture (because these interfaces are not yet standard), and thus by being present at various levels of the value chain, from components to complete systems.
>
> But once most customers are satisfied with the performance of a product on this dimension (i.e., product reliability is 'good enough'), the basis of competition changes. Making even more reliable products no longer yields superior profits. At that point, the best way to create more value is to improve the product on new performance dimensions valued by customers, such as time-to-market (speed), convenience, price, serviceability, etc.
>
> In this context, non-integrated specialists, with less at stake in the old model, are better positioned to meet customers' needs. As more of these players provide specific pieces of value-added activity and as interfaces within products get more standardized, the products themselves become more modular and the value chain 'dis-integrates'.
>
> At some later point though, changes in the underlying technology (with implications on product architecture), or the emergence of a new functionality, or dynamics of commoditization in some parts of the value chain, will lead players to re-integrate to regain some competitive advantage, and the cycle will start again.

THE BIG WHITE SPACE OF THE EUROPEAN A&D INDUSTRY

MAY 2013

THE ABORTED MERGER OF EADS and BAE Systems last year reignited talks about a long-awaited wave of consolidation among prime aerospace and defense (A&D) contractors. After all, Europe still has four major shipbuilders, four missile manufacturers and three combat aircraft primes, while the larger US defense market has essentially two players in each of those categories. The speculations were underpinned by a basic but deeply erroneous assumption: "Bigger is better."

In Europe, the top five industry players— EADS, BAE Systems, Finmeccanica, Safran and Thales—are big enough, with more than $15 billion each in annual A&D revenue. If anything, they are too big for their own sake. What is needed is not further consolidation but rather a reshuffle of their business portfolios so each of these groups is worth more than the sum of its parts. By and large, that is not the case today. Because of a deep-rooted vertical integration and a predatory mindset, each of these companies is a mixed bag of multiple businesses, many of which would be better off on their own or under smaller companies.

Apart from large passenger aircraft manufacturers—Airbus and Boeing—no part of the A&D business requires a company to have more than $10 billion in annual sales to succeed. You do not need to be that big to compete: $4-8 billion in revenue to be

The predatory behavior of Europe's top players has prevented the emergence of mid-sized merchant suppliers with critical mass in such sectors as defense electronics or aerostructures.

an original equipment manufacturer of business aircraft (Gulfstream and Dassault) or regional aircraft (Embraer and Bombardier); $4 billion to be a world leader in tactical missiles (MBDA) or to design nuclear submarines and warships (DCNS). Dassault Aviation may be "just" an 11,000-employee company with $5 billion in sales, yet it profitably makes and sells some of the most advanced combat aircraft and business jets. The company has established a strong market position without any large M&A deal or adventurous build-up.

In fact, bigger is often worse. Beyond a certain threshold of size and diversification, dis-economies of scale are bound to appear, typically due to the increased complexity of accommodating different business models and integrating new acquisitions. Some companies have succeeded by managing their businesses as loose portfolios of independent, relatively specialized operations. But even they have limits. L-3 Communications thrived under such a model until its portfolio of dozens of loosely managed businesses grew so big that it became unwieldy. L-3's response was to move toward becoming a more traditional, vertical organization. As soon as it changed its model, L-3's performance started deteriorating.

Top European players should aim to manage their portfolios more dynamically, making sure all their businesses gain from their ownership.

One of the oft-discussed benefits of an EADS-BAE tie-up was that it would have created the world's largest aerospace company, with annual sales approaching $100 billion. However, the weak link in Europe's A&D industry is not at the top, but rather in the middle of the pyramid, where there is what I call a "big white space" (*see chart below*). The predatory behavior of Europe's top players has prevented the emergence of mid-sized merchant suppliers with critical mass in such sectors as defense electronics or aerostructures. As a result, there are essentially no European $5-15 billion A&D suppliers.

THE BIG WHITE SPACE OF THE EUROPEAN A&D INDUSTRY
(PLAYERS WITH OVER $1Bn of A&D REVENUES)

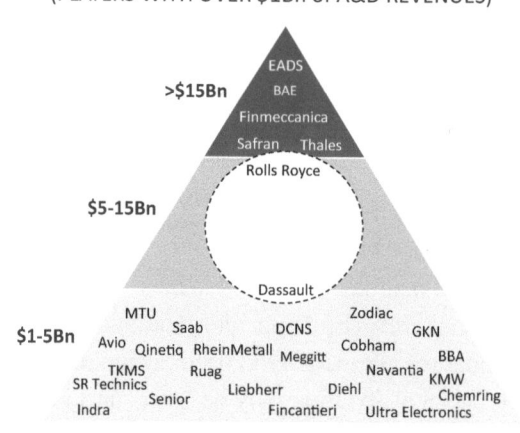

>$15Bn: EADS, BAE, Finmeccanica, Safran, Thales

$5-15Bn: Rolls Royce, Dassault

$1-5Bn: MTU, Saab, DCNS, Zodiac, Avio, Qinetiq, RheinMetall, Meggitt, Cobham, GKN, BBA, TKMS, Ruag, Navantia, KMW, SR Technics, Senior, Liebherr, Diehl, Chemring, Indra, Fincantieri, Ultra Electronics

Sagem was a rapidly growing, $4 billion communications and defense electronics group before Snecma (now Safran) scooped it up in 2005. Since then, all Sagem's businesses have been disposed of except for Sagem Defense, which has been marginalized. Had it remained independent, Sagem would have been a perfect platform to create a leading defense electronics supplier in Europe. Today, the largest such company is the $1.2 billion Ultra Electronics.

So instead of trying to become bigger, top European players should aim to manage their portfolios more dynamically, making sure all their businesses gain from their ownership, spinning out those that do not and letting them become growth platforms for smaller players. That would be much more beneficial to all their European stakeholders. Rather than wondering whether there are too many primes in Europe and whether they are big enough, one should worry about how many mid-size suppliers Europe does not have and create the conditions for such players to emerge. The future of the European A&D industry is at stake.

KEY CONCEPT

Economies vs. Dis-Economies of Scale

Economies of scale occur when the average cost of producing a product or service declines as the product or service is produced in larger volumes. This is essentially the case when the underlying cost structure is predominantly fixed (i.e., independent of the volume produced). As larger volumes are produced, those fixed costs can be amortized on a larger base and thus unit costs decrease accordingly.

By extension, it is generally assumed that, as a firm grows in size, it can reap economies of scale as long as its fixed cost base increases less rapidly than its revenues. This is the case, for example, if one can double or triple production by adding shifts to an existing machine or factory, instead of adding machines or factories. Thus, the traditional scale curve looks as follows:

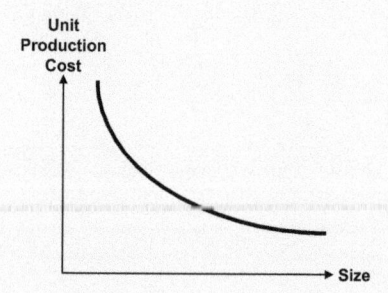

However, beyond a certain size, economies of scale tend to flatten and possibly even turn to dis-economies of scale. The main driver of dis-economies of scale is complexity. As the organization gets bigger and more complex, specific resources – typically corporate overheads – are required just to manage the complexity of the business (e.g., managing multiple businesses with different business models in multiple countries, selling multiple products to multiple customers, etc.). Thus, the scale curve may end up looking more as follows:

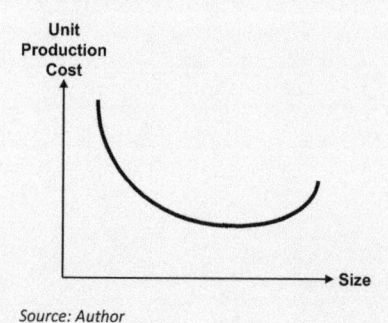

Source: Author

One way to avoid dis-economies of scale is to break up the organization into smaller, more autonomous business units, and to streamline corporate overhead functions.

INDUSTRY STRUCTURE AND COMPETITION

THE NEW PREDATORS

MARCH 2020

BEHIND THE BIG AEROSPACE and defense (A&D) primes like Boeing, Airbus or Lockheed Martin, and the "Super Tier-1s" such as United Technologies (UTC), GE or Safran, a quite different type of company is shaping the global A&D industrial landscape in a way that may be even more impactful than high-profile UTC-Raytheon-type mergers.

Companies such as Teledyne Technologies, TransDigm Group and Heico Corporation are the spearheads of a breed of A&D players dedicated to "components and subsystems," with explicit and perfectly executed "horizontal" external growth strategies. Their track records are impressive: These three companies—with combined revenues of more than $10 billion—have collectively made close to 200 acquisitions and delivered more than 20% average annual growth rate in either profitability or share value over the last 20 years. Thanks to such returns and skyrocketing market valuations, they are able to outbid most other contenders when going after an acquisition target by leveraging the so-called "accretive effect", which boosts the acquiring company's earnings per share, as long as the price paid for the target as a ratio of the enterprise value (EV) over its earnings before interest, taxes, depreciation and amortization (EBITDA) is lower than that of the acquiring firm. As it happens, the current EV/EBITDA ratio of the three above-mentioned companies stands at more than 18. By comparison, most other A&D companies have an EV/EBITDA ratio in the 9-13 range (*see graph below*).

> Companies such as Teledyne Technologies, TransDigm Group and Heico Corporation are the spearheads of a breed of A&D players dedicated to "components and subsystems," with explicit and perfectly executed "horizontal" external growth strategies.

Such "buying power" is enhanced by economies of scope and operational synergies (in corporate overheads, sales and marketing, etc.), which immediately boost the profitability of the acquired company and can therefore be factored in the offer price. This gives them an additional edge against pure financial investors like private equity (PE) funds, which have historically been strong buyers of such component and subsystem businesses.

Two recent deals in Europe (one still ongoing) illustrate this new balance of power. The first concerns Souriau-Sunbank, a $360 million-revenue specialist in interconnection technology for harsh environments. After being owned successively by two PE funds and bought by Esterline (now TransDigm) in 2011, it was again put up for sale last year. While expectations were that a PE fund would grab it, another industrial buyer, Eaton Corp., won the contest, paying the hefty price of $920 million (an EV/EBITDA multiple of 12). The second deal relates to a French company called Photonis, a world leader in night-vision technology for defense and space applications, for which Teledyne is apparently bidding—and offering a price 30% higher than the highest PE bid!

EV/EBITDA RATIO OF SELECTED A&D COMPANIES
(as of 02/26/2020)

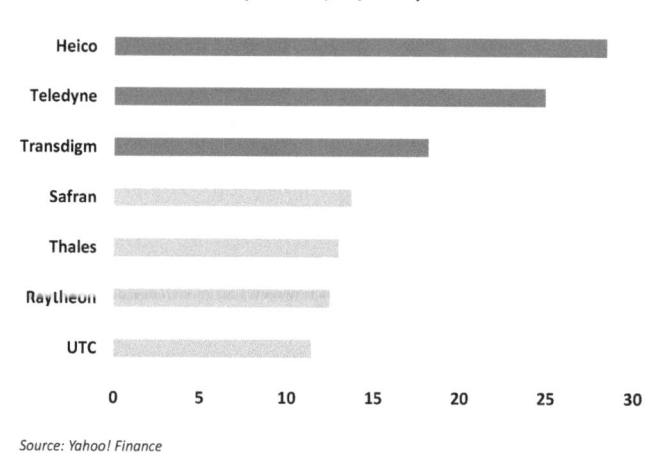

Company	EV/EBITDA
Heico	~30
Teledyne	~25
Transdigm	~20
Safran	~12
Thales	~11
Raytheon	~10
UTC	~9

Source: Yahoo! Finance

> By combining "private equity-like growth in value with liquidity of a public market," they are not only beating PE players at their own game, but they are also capturing a significant share of the A&D capital market by offering investors an attractive alternative to the traditional large "Tier-1s".

These deals highlight the limits of the traditional private equity model (too short-term and too short-sighted) and why the "new predators"—all publicly listed companies—are in a much better position to continue to thrive. In fact, by combining *"private equity-like growth in value with liquidity of a public market,"* as TransDigm puts it, they are not only beating PE players at their own game, but they are also capturing a significant share of the A&D capital market by offering investors an attractive alternative to the traditional large tier-1s such as UTC, Thales or Safran. These groups are typically too busy focusing on large systems and equipment to realize that they would benefit from articulating a proper "component and subsystem" strategy. They would benefit not only because their portfolios are still full of such businesses, but also because their long-term competitiveness largely depends on their ability to nurture a strong network of strategic suppliers, in terms of criticality to their own systems as well as national sovereignty.

As it happens, Photonis seems to be such a strategic supplier, since the French government recently announced it would veto the Teledyne deal, hoping to give other French or European companies or investors time to make a competitive offer for the business. But because PE funds, at least in Europe, are somewhat faint-hearted when it comes to ambitious sector-specific "horizontal" portfolio strategies, and because Europe has no industrial player able to compete with the likes of Teledyne, the outcome of the process is still highly uncertain.

In the meantime, it remains that Teledyne, Heico, TransDigm and the like are surreptitiously reshaping the A&D industrial landscape by buying technological nuggets and component businesses left and right, on both sides of the Atlantic. In the process, they are boosting their shareholders' returns and changing the balance of power with both traditional private equity investors and large, vertically integrated A&D groups. As the saying goes: One man's meat is another man's poison.

KEY CONCEPT

THE ACCRETIVE EFFECT

A merger and acquisition (M&A) deal is said to be 'accretive' if the price paid by the acquirer for the target as a ratio of the enterprise value (EV) over its earnings (also called price-earnings- P/E - ratio) is lower than that of the acquiring firm. In that case, it is expected that the Earnings per Share (EPS) of the acquirer will increase after the deal. In the opposite case (EPS expected to decrease), the deal is said to be 'dilutive'.

Even though it does not reflect the real value potentially created by an acquisition, the accretion/dilution test is a basic way to justify the "price" paid for an acquisition by making the financials work at a high level.

In practice, value in an accretive acquisition is generated when the buyer of a smaller (and cheaper) company is able to bring the profitability and value of the acquired company to the level of its own profitability and value, either by generating economies of scale or scope (e.g., in overhead or R&D costs) or by adapting the acquired company's business model to its own to make it more profitable (for example by improving business processes such as tender bidding, product development, etc.).

A AND D DIVERGENCE

AUGUST 2015

AEROSPACE AND DEFENSE (A&D) has historically been an industry that encompasses manufacturing of aircraft, ships, weapons, space rockets, satellites and military land vehicles. This is rooted in the days before commercial aviation existed, and aerospace companies were focused on the governmental military market.

Indeed, in the 1960s, the top "armament" manufacturers were all military aircraft manufacturers: Boeing, Lockheed Corp., General Dynamics, United Aircraft, the Martin Co., Douglas Corp., Hughes Aircraft Co. and McDonnell Aircraft. But today, with the rapid growth of commercial aviation, the label "aerospace and defense" has become largely irrelevant, if not entirely misleading.

For a start, the non-aerospace defense sector is dwarfed by the overall aerospace sector to the point that it has become marginal for players that are still involved in both. The non-aerospace defense sector represents less than 15% of the turnover of Boeing and the Airbus Group, for example. Within the aerospace sector itself, the military aircraft business has become increasingly small, representing less than 20% of aerospace revenues of the likes of Boeing, Airbus, GE Aviation and Safran (*see graph below*).

Airbus makes money delivering airplanes to airlines; Lockheed Martin makes money doing business with the Pentagon.

Moreover, while the defense and space parts of aerospace have had much in common (hence the rationale for creating "defense and space" divisions within companies), the space sector is undergoing a revolution that is bringing it closer to the commercial aerospace sector. As an illustration, when bidding for the OneWeb Internet satellite constellation project this year, Airbus emphasized as a competitive advantage its experience mass-producing tens of airplanes monthly. Even the launcher business, traditionally linked to the defense business, is being reinvented by new players like SpaceX.

Beyond the size difference and evolution of the space sector, the very nature of the defense business has become so distinct that a company can no longer define itself as being both an "aerospace" and "defense" firm. There are aerospace companies doing business in defense, and there are defense companies offering aerospace products. This is not just a play on words— the differences in underlying economics and capabilities are huge. Airbus Group is an aerospace company, not a defense company. Lockheed Martin is a defense company, not an aerospace company. That is why Airbus can churn out 42 A320s a month while Lockheed is running billions of dollars and years behind on the F-35 Joint Strike Fighter program. Airbus makes money delivering airplanes to airlines; Lockheed Martin makes money doing business with the Pentagon. This is not the same business.

BOEING, AIRBUS & LOCKHEED MARTIN 2014 REVENUES BREAKDOWN

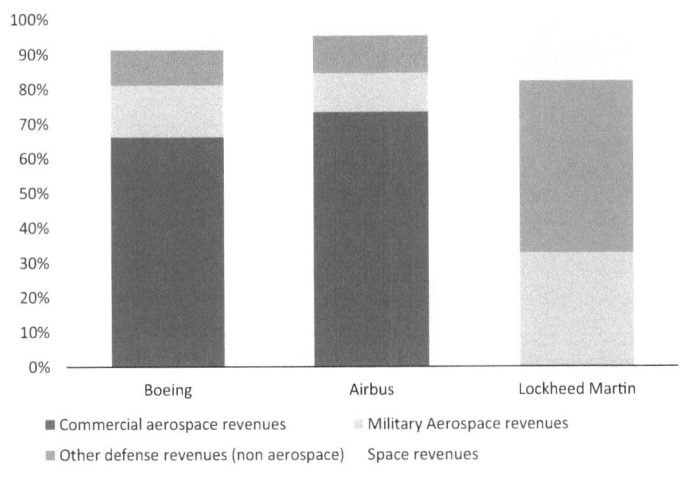

- Commercial aerospace revenues
- Military Aerospace revenues
- Other defense revenues (non aerospace)
- Space revenues

Source: Company Annual Reports

Of course, one of the strongest myths tying together aerospace and defense is the counter-cyclicality of the two sectors. But, beyond a certain size, any business has to be self-reliant and able to cope with down cycles. As a matter of fact, two-thirds of the world's top 100 A&D companies are "pure or semi-pure plays," with at least 75% of their business either in commercial aerospace or in defense. Over the last decade, most defense companies have been diversifying not by getting into commercial aerospace but by developing their security and IT business to the point that it has become more common to use "defense and security" than "aerospace and defense" to define their industry sector.

There was a time, though, when defense products—missiles and fighter aircraft in particular—were the flagship of aerospace company offerings. It was clearly the case in the '80s and early '90s. Innovation,

> *Two-thirds of the world's top 100 A&D companies are "pure or semi-pure plays," with at least 75% of their business either in commercial aerospace or in defense.*

technology and money was flowing in from the defense side into the commercial side of the business. Now it is the other way around. Defense companies have become relative laggards in innovation and their market has been shrinking considerably.

Defense companies are indeed in an uncomfortable position, being challenged on the one hand by changing customer needs and on the other by new competition from companies outside the traditional military-industrial complex. We should therefore stop wondering why aerospace companies exit or do not invest in the defense sector. Rather, we should wonder why some firms still pretend (or believe) that they are both aerospace and defense companies when in reality the two are increasingly at odds.

KEY CONCEPT

COUNTER-CYCLICALITY

The counter-cyclicality argument relies on the fact that defense cycles have historically been much longer than commercial aerospace cycles: 13-21 years with nine years between inflection points for defense vs. 10 years with four years between inflection points for commercial aerospace. Thus, it is very unlikely that both sectors will be in a down cycle at the same time.

But the argument also relies on the assumption that there are enough synergies between both sectors (typically in terms of R&D and manufacturing resources) to be able to transfer workload from one to the other if necessary.

The smaller the company, the more likely it is to use shared resources and therefore the more it will benefit from having a somewhat balanced portfolio between commercial aerospace and defense activities, in case of a down cycle.

The situation is different for a large, multi-billion firm. One should expect each part of the business to have a critical size and any area of shared resources to be proportionally small. Therefore, resilience – when needed – must be found within the business, not outside it.

So, even if the statement that *"beyond a certain size, any business has to be self-reliant and able to cope with down cycles"* is being severely tested by the current crisis, the counter-cyclicality argument will always be stronger for a small company, if only because, before being resilient, one must first survive. And the survival threshold will always be much lower for a $100 million entreprise than for a multi-billion corporation.

DIGITAL AEROSPACE

APRIL 2015

THE ERA OF DIGITAL AEROSPACE is upon us. From design offices to flight decks, from clean rooms to command-and-control rooms, the digital economy is seeping through every corner of the aerospace and defense (A&D) industry, and it is just the beginning.

In this process, the industry is being turned upside down. What used to be predominantly defined by hardware is now increasingly ruled by software. Indeed, that is how the latest generation of A&D products such as radio communications systems or satellites are named: They are "software defined" and can upgrade and reconfigure themselves.

These products are connected within networks that are themselves self-organizing and self-healing, making the whole and the parts "future-proof" against changing standards and emerging applications. Everything is controlled through software, to the point where the capabilities of a given product are more limited by its digital library than its physical characteristics.

Similarly, in commercial aviation, inflight connectivity is defining cabin interiors, while data analytics is transforming flight management and aircraft maintenance. Air traffic management is also undergoing a major software-driven revolution, with the emergence of remote and virtual control towers. A data network transfers high-definition images and all relevant airport systems to an integrated controller station that can be pretty much anywhere.

The next frontier is clearly going to be remotely piloted aircraft systems. When this happens, who will be at the forefront of the industry? Will Boeing, Airbus, Lockheed Martin and the like still be running the show?

Interestingly, today there is not a single discussion about the future of aerospace that does not mention companies such as Google, Amazon, or Apple. Why? Because these players are digital natives, so to speak; it is in their DNA to figure out how to create value in the digital economy, regardless of the medium. It does not matter whether it is a phone, a car, a home or—ultimately—an airplane. These are just nodes in a giant digital network. Ten or 15 years ago, if

The digitalization of the economy is blurring traditional boundaries between sectors, creating opportunities for new entrants and technology spill-ins from multiple directions.

someone had predicted that Google, Apple or an Internet entrepreneur named Elon Musk would make cars, nobody would have taken them seriously. Yet, it is happening.

But what exactly is happening? First, the digitalization of the economy is blurring traditional boundaries between sectors, creating opportunities for new entrants and technology spill-ins from multiple directions. Second, as the digital economy rapidly expands, the value-creation events—or economic touchpoints—are moving away from the players that are predominantly involved with hardware or platforms, which is still the case of most A&D companies.

Therefore, these companies are at risk of becoming marginalized and their products ultimately commoditized. That is why, in spite of comparable revenue and employee numbers, the market caps of Google, Amazon, Apple etc. are several times higher than those of traditional A&D companies such as Boeing and Airbus (*see graph below*). This value gap is a reflection of the perceived vulnerability of "old economy" players to major disruptions, be it new competitors, new technologies or new applications.

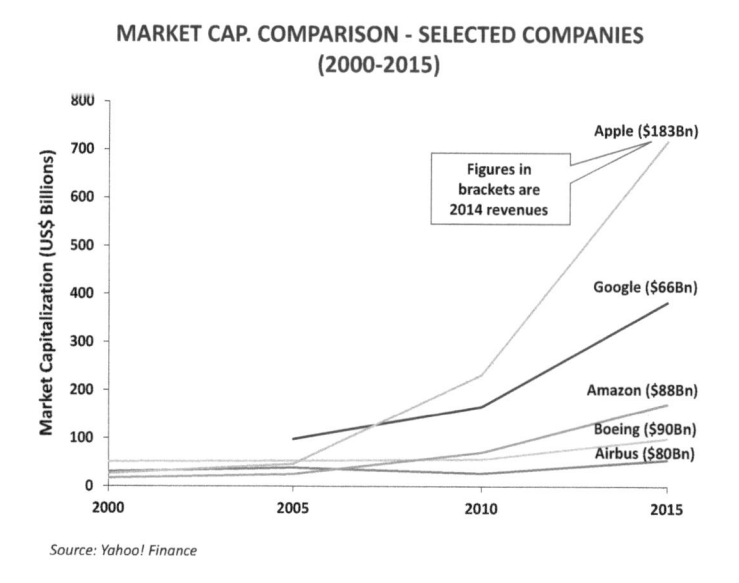

MARKET CAP. COMPARISON - SELECTED COMPANIES (2000-2015)

Figures in brackets are 2014 revenues

Apple ($183Bn)
Google ($66Bn)
Amazon ($88Bn)
Boeing ($90Bn)
Airbus ($80Bn)

Source: Yahoo! Finance

To see such vulnerability, one just needs to look at the philosophy behind the new economy's software development culture. This was captured in 2001 by a group of software developers in a statement called the "Manifesto for Agile Software Development[i]." It essentially says improvements in software development will be achieved by *"valuing individuals and interactions over processes and tools, working software over comprehensive documentation,*

---

[i] **Manifesto for Agile Software Development:** https://agilemanifesto.org/

customer collaboration over contract negotiation and responding to change over following a plan." How much further could it be from the A&D industry's traditional work practices?

To remain at the forefront of their industry, A&D players therefore don't just need an evolution; they need a revolution. Instead of looking at their environment from an inside-out perspective— this is where we are, these are our options to grow and improve—they need to go through an "out-of-body" experience and look at their environment (and themselves) from an outside-in perspective, rethinking their business boundaries and purpose in the process.

By doing so, they will see a completely different picture: one where R&D is financed by venture capitalists and web entrepreneurs instead of institutional investors or industry heavyweights, where video game algorithms are more complex than spacecraft's, where commercial technology outperforms military technology, and where collective ingenuity and individual talent trump bureaucracy and managerial processes.

> *As the digital economy rapidly expands, the value-creation events—or economic touchpoints—are moving away from the players that are predominantly involved with hardware or platforms.*

Mission impossible? Probably, unless visionary leaders emerge and show the way. Until then, get used to seeing aerospace engineers being displaced by video game developers.

KEY CONCEPT

ECONOMIC TOUCHPOINTS

Economic touchpoints are interface points between consumers (retail customers) and a brand, product, or service, where economic value is created, that is where a product or service is actually purchased. From there on, economic value trickles down the value chain to all the players that have contributed to bringing the product or service to market.

In the digital economy, these economic touchpoints are moving away from suppliers of physical products and even from physical channels. Instead, they are increasingly controlled by pure digital players such as Amazon, Google, Uber, etc. As the 'digital pie' keeps growing, so do the revenues of these 'digital portals'. Meanwhile, more traditional (hardware) players are sliding down the value chain, further away from the economic touchpoints and value creation events. In such context, their added value will become much less visible and their "share of the pie" is likely to keep getting proportionally smaller, even if it grows in absolute value: out of sight, out of mind…

Aerostructures: The Next Battleground

April 2018

MOST CURRENT DISCUSSIONS ABOUT commercial aircraft innovation center on areas such as engines, power systems (with the electric aircraft in the line of sight), connectivity and Big Data. But while these areas of innovation are relevant for the relatively few players involved in them, the stakes are much bigger in what could be considered the "entrails" of the aircraft: the aerostructures, materials and detailed parts, which concern a much bigger pool of players and are the largest contributor to an aircraft's cost.

Indeed, excluding powerplants, the airframe and all its constituent parts account for two thirds of a commercial aircraft's costs. And two thirds of all companies involved in the making of an aircraft produce materials, detailed parts and/or aerostructures, while hundreds of subcontractors still make some of the millions of parts that make up the airframe.

Despite these telling numbers, aerostructures are not typically thought of as the next aerospace battleground because innovations have so far been slow and relatively few and far between. The most significant has been the introduction of composite materials, which have taken an increasing share of the flyaway weight, reaching 50% in the Boeing B787 and Airbus A350. Yet it took 30 years for the use of CFRP (Carbon Fiber Reinforced Polymer) to move from minor sub-assemblies such as fins and rudders to the full fuselage section. And composites still account for only 5% of the total aircraft raw material demand (in weight), which has so far limited their disruptive impact.

But it looks increasingly likely that aerostructures will determine the winners and losers in the next development phase of the commercial aerospace industry, for the following reasons.

> *It looks increasingly likely that aerostructures will determine the winners and losers in the next development phase of the commercial aerospace industry.*

Technologically, ongoing advances in both metal alloys (such as titanium aluminide or aluminum-magnesium-scandium) and thermoplastics, combined with rapidly maturing manufacturing techniques such as 3D printing, robotics, out-of-autoclave processing and fusion welding are generating a step change in possibilities (and expectations) for weight and cost savings, as well as lead times on airframe structures and parts.

While each type of innovation can individually be perceived as incremental, together their potential impact is truly revolutionary. Suddenly millions of fasteners and thousands of processing hours can realistically be eliminated, and the prospect of a "buy-to-fly" ratio (the weight of raw materials required over the weight of the component itself) close to 1:1 becomes achievable (from a current average level of 10:1).

Industrially, this inflow of new technologies will force suppliers to make major investment decisions that will favor either technology specialists with a unique depth of expertise or large players able to offer a broad panel of capabilities, across different materials and manufacturing techniques. In that process, it is likely that many incumbent suppliers probably will end up "stuck in the middle", with neither the depth nor the breadth of expertise required to remain competitive.

Commercially, as it becomes increasingly difficult for Boeing and Airbus to differentiate their products based on systems and equipment - which are essentially made by the same global suppliers - the only area where they will still be able to distinguish their products will be the overall airframe design and production. These will entail lighter, more integrated and robust structures as well as more cost-efficient assembly processes and ultimately cheaper products. It is therefore no surprise that both airframers are investing huge amounts in optimizing and transforming their aerostructure capabilities and supply chain.

Finally, aerostructures are the only area where Original Equipment Manufacturers (OEMs) can realistically integrate suppliers from strategic countries such as China, India, or Russia into their supply chain. While these countries almost totally rely on foreign technology for systems and equipment, they already have some significant domestic capabilities in materials and aerostructures as well as huge resources and ambitions to beef them up, making their integration in Airbus' and Boeing's global supply chain highly probable.

All these elements are bound to lead to a complete reshuffling of the aerostructure and detailed-parts landscape, forcing established players to radically transform their business.

What used to be perceived as an unavoidable trade-off between profitability and growth will become a trade-on opportunity combining innovation and improved cost performance.

Now, it is not enough for them to deliver on time and match their customers' incremental cost-reduction demands. They need to move to a new cost curve altogether (*see graphic below*), which can only be achieved by making drastic strategic investments in new technologies or in external growth. By doing so, what used to be perceived as an unavoidable trade-off between profitability and growth will become a trade-on opportunity combining innovation and improved cost performance. As one industry executive told me, "One cannot keep shifting from one priority to another, from achieving technical excellence to reducing costs. Instead, one should aim at addressing those in parallel, in such a way that they reinforce each other."

KEY CONCEPT

Cost Curves

In general, competitors in the same industry share a similar cost curve (A) and the name of the game is to go as far down the curve as possible, as quickly as possible. This assumes that the underlying technology (broadly speaking) remains constant. When the underlying technology radically changes, it potentially defines a new industry cost curve (B), which is below the existing cost curve A. As such, it resets the economics of the business altogether.

While initially, early adopters of the new technology may be less cost competitive than suppliers sticking with the old technology (due to experience and scale curve effects), ultimately, they are expected to become much more competitive as they move down the new cost curve B. Therefore, to remain competitive, all suppliers will need to move from cost curve A to curve B, which can only happen if they switch to the new technology.

This is what is likely to happen in Aerostructures due to the combined impact of new material technology and new manufacturing technologies such as 3D printing, robotics, out-of-autoclave processing, fusion welding, etc., which, together, have the potential to bring about a completely new cost curve.

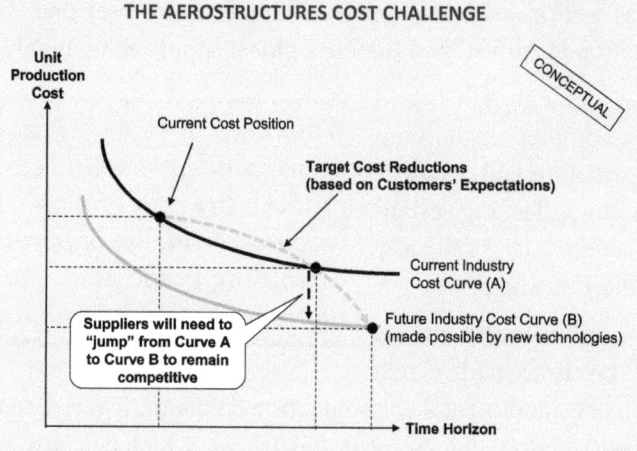

THE AEROSTRUCTURES COST CHALLENGE

Source: Author

A Duopoly on the Edge

December 2020

THE BOEING-AIRBUS RIVALRY in commercial aerospace has become so engrained in popular culture that it would be easy to forget that the duopoly truly started just two decades ago, after Boeing acquired McDonnell Douglas in 1997. Since then, the two companies have been fighting tooth and nail to launch new programs, win customer orders and ramp up production rates.

Both companies reached their peak in 2018, collectively and evenly delivering 1,600 aircraft and generating $120 billion of revenues. At that point, they were both anticipating glorious days ahead, predicting a 20-year demand for 40,000 new aircraft worth $6 trillion, that they would happily share between them.

However, as both companies were celebrating their record deliveries, alarm bells started ringing. Two back-to-back crashes brought the whole Boeing 737MAX fleet to the ground. Stories about recurring quality issues with the 787 Dreamliner also emerged. At the same time, Airbus announced that it would stop producing its A380 jumbo jet after selling only 250 units of an aircraft that cost more than $15 billion to develop. Meanwhile, orders for its most recent wide-body A350 were drying up with less than 100 new orders in three years, forcing the company to reduce its target annual production rate for 2020 from 140 in 2017 to 104 in February 2020, just before the COVID-19 crisis struck. Since then, of course, things have gotten much worse and both companies have been scrambling to mitigate the crisis's damages and reset their growth trajectory and goals for 2021 and beyond.

In the wake of such unprecedented shock, it is worth taking stock of where the Boeing-Airbus duopoly stands and what the future holds for both players.

At first glance, Boeing seems to be in a much worse situation than Airbus, with its 737MAX not yet back in service and the 787 issues not solved. In 2020, Boeing only delivered 157 aircraft for the year, while Airbus has managed to deliver 566 aircraft, 34% less than in 2019 but enough to give the European company an unprecedented 78% market share for yearly deliveries (*see first graph below*). Airbus' order book is now almost 70% larger than Boeing's. Additionally, Boeing has spent the last two years essentially firefighting, which means it has probably lost some ground in terms of preparing for the next generation of airplanes. Meanwhile, Airbus has been carrying on with the development of its A321XLR (extra-long range) aircraft, which not only allows the European company to sustain its dynamics of continuous innovation but also potentially pulls the rug from under Boeing's feet for addressing the "middle of the market" segment due to its advantageous range-payload characteristics.

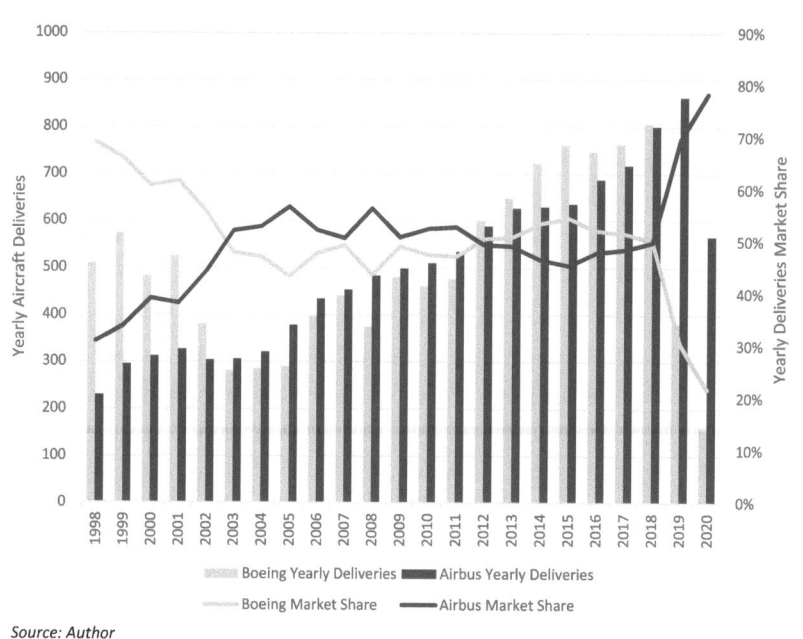

BOEING-AIRBUS YEARLY DELIVERIES AND MARKET SHARES 1998 - 2020

Source: Author

Nevertheless, one should not assess Airbus' strengths simply based on Boeing's current weaknesses. Airbus has also suffered from the crisis and in some respect is still structurally weaker than Boeing, first because its commercial aerospace business pretty much stands on its own (whereas Boeing can rely on a large and relatively stable defense and space business), and second because its industrial organization remains a multinational puzzle that is far from being optimized and still subject to political tensions between its home countries.

It will have taken Airbus roughly 50 years to overcome Boeing as the leading commercial aircraft supplier in the world.

All things considered though, Airbus has never been in a better position to take a significant lead over Boeing. Based on their respective pre-COVID growth rate, by 2027 (and possibly earlier), Airbus will have delivered as many airplanes as Boeing since its first official airplane delivery in 1974 (*see graph below*). This means that it will have taken Airbus roughly 50 years to overcome Boeing as the leading commercial aircraft supplier in the world.

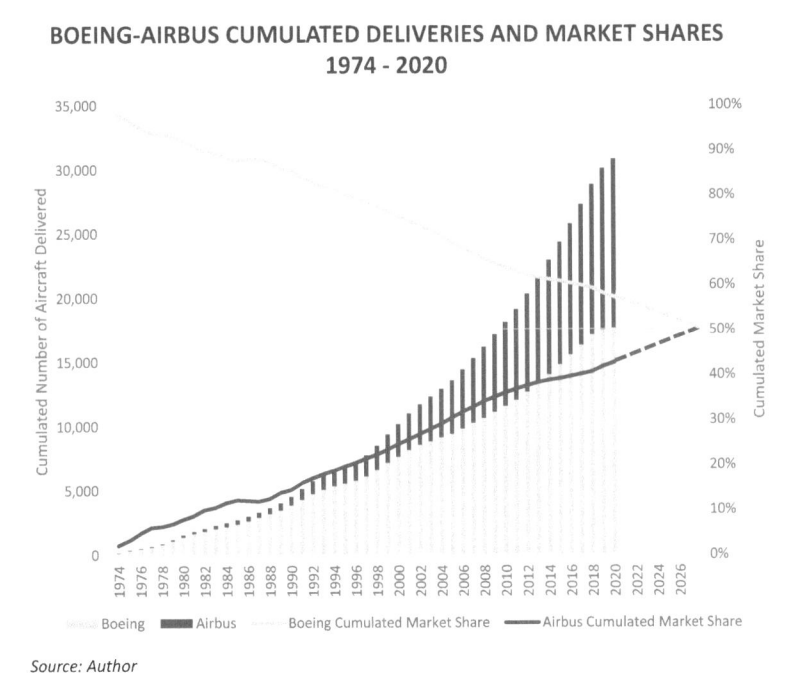

BOEING-AIRBUS CUMULATED DELIVERIES AND MARKET SHARES 1974 - 2020

Source: Author

But even if the balance of power lastingly tilts towards Airbus, it should not fundamentally alter the duopoly dynamics in the medium term. The two players will still be dominant and compete head-to-head. Beyond the current decade though, the big question is what the impact of China's forthcoming entry will be.

Indeed, China's first indigenously developed large commercial airplane – COMAC's narrow-body C919 – is expected to come into service within a couple of years. This will mark the country's official entry into the large commercial aircraft market and a major milestone of a long-term strategic play underpinning colossal geopolitical, industrial, and commercial stakes. To succeed, it will be able to rely on a huge, at least partly captive, domestic market, as well as on an aggressive pricing strategy (it is expected that the C919 could be priced 50% lower than the A320 or 737). But above all, selling commercial airplanes will become part of its "soft power" strategy.

And while it will take some time before European or American airlines buy a Chinese airplane, airlines in other regions may be more easily enticed by the Chinese products. Newly rich countries like India, Indonesia, Turkey,

Russia, Egypt – all of which will be among the world's wealthiest nations by 2030 – as well as increasingly populated regions like East and West Africa (which will account for most of the world's demographic growth for the rest of the century) will all be addressable markets for the C919 and its successors.

Of course, there are still many hurdles to be overcome for COMAC to become a fully-fledged competitor to Airbus and Boeing – not the least developing a robust supply chain and an international sales and service network – but in a world where trade wars could become increasingly the norm, China has a lot of assets to put forth.

China's entry is bound to alter the economics and the politics of the business in the long term.

Whereas the Covid-19 crisis has significantly dented both Airbus and Boeing's growth trajectories, it has boosted China's quest to become an aerospace powerhouse, in part thanks to a domestic air travel market more buoyant than ever (back to its pre-COVID level and on the verge of becoming the largest in the world). And while it is unlikely the Boeing-Airbus duopoly will disappear any time soon, China's market entry is bound to alter the economics and the politics of the business in the long term, if only because it will force Boeing and Airbus to rethink their global industrial strategy and to take more risks in terms of new product development.

Contrary to appearances, the Airbus-Boeing duopoly has never been a long, quiet river. It is certainly far from being one right now and both players should brace themselves for strong currents and treacherous rapids ahead.

KEY CONCEPT

Duopoly

A duopoly is a competitive situation where two players are dominant to the point of controlling the market and new entrants. A duopoly is more likely to occur when the following conditions are met:

- High fixed costs (thus high minimum efficient scale) and a steep experience curve, both generating high entry barriers
- A fragmented customer base, tilting the bargaining power towards the suppliers
- High switching costs for customers (due to brand loyalty, product dependencies, training requirements, etc.)
- Common standards (or agreement to drive standards in the same direction), thus raising the regulatory bar for new entrants
- Highly integrated value chain (either through proprietary architecture or risk-sharing partnerships)

More specifically, the Boeing – Airbus duopoly has been able to perdure for four main reasons:

- **Very high "cost base"** compared to the size of the market, making the investment unattractive for a rational economic player looking for a reasonably certain return on investment. As it is, the market is barely big enough for two players to be profitable.
- An **incompressible time to market**, driven in part by stringent regulatory requirements. A new entrant would face a long and uncertain timeframe to envisage a market entry at scale (unless it has a privileged access to a large enough launch market, as should be the case for the Chinese).
- An **incremental innovation strategy** by Boeing and Airbus, that has created multiple dependencies between (generations of) products, thus leading to **high switching costs** for customers (if they were to take on a completely new type of airplane).
- The industry's **economic benefits are mainly indirect**: skilled labour, industrial know-how, trade balance, and political soft power. Thus, only the largest and richest economies (namely the US, the EU and now China) can afford and justify financing it. Even Japan and Canada – two of the world's richest nations – have not been able to succeed on their own.

Chapter 2

SUPPLY CHAIN DYNAMICS

"There is a kind of success that is indistinguishable from panic."

Edgar Degas, French artist (1834–1917)

CHAPTER INTRODUCTION

This chapter looks at the industry dynamics from the point of view of the A&D supply chain, which is made up of thousands of small and medium-sized – often captive – subcontractors or suppliers of parts and sub-systems, in contrast with their relatively few and large customers: OEMs (Original Equipment Manufacturers), prime contractors and system integrators.

The A&D supply chain is thus overly complex, with so many players involved in making millions of parts on all continents. It is also characterized by a high level of interdependency between suppliers and customers, the former being often highly dependent on one or two big customers and the later often relying on suppliers as single sources for specific products.

The chapter's first two columns highlight the challenges traditional sub-contractors – especially in Europe – face in keeping up with their customers' always more stringent requirements and in remaining competitive at a time when the need for capital investment and internationalization take them far away from their comfort zone. Key challenges typically involve the ability to raise capital and to ramp up production, while maintaining high operational performance standards and expanding their industrial footprint to "low-cost countries" to remain competitive.

The last two columns take a more systemic view, first by analyzing the drivers of competitiveness for the supply chain as a whole – from the point of view of large OEMs such as Airbus or Boeing – and then by proposing a new, more sustainable approach to customer-supplier relationship management to address the lack of communication and mutual trust still pervasive in the A&D supply chain.

Time to Bridge the Industry – Finance Gap

December 2013

THE WORLDS OF FINANCE AND INDUSTRY do not often mesh well. This disconnect is especially true in the aerospace and defense (A&D) industry, which is dominated by engineers who believe in the value of "making things" as opposed to just "making money." The root of the problem is that the two sides simply do not know or understand each other well. A&D companies have historically relied on their customers more than bankers for funding. To some of them, customer-funded R&D is a more familiar term than financial debt. Many smaller businesses do not even have a proper CFO.

I recently worked on a transaction where a long-time entrepreneur had finally agreed to sell his company – a small supplier of aerospace sub-systems, provided it would be bought by a wealthy fellow entrepreneur, worthy of his trust and status. I suppose he expected to be handed a big personal check. But when he realized that there would be some debt and financial leverage involved, he became uncomfortable and backed out of the sale.

This case may be extreme, but the reality is a lot of A&D companies remain suspicious of financial investors such as private-equity or hedge funds. Unfortunately, such suspicion is not completely unfounded.

Private equity is like cholesterol: There is a good type and a bad type. The latter will have a purely financial engineering approach to a deal: Leverage the purchase, milk the company to repay the debt quickly and flip it to the next buyer, with complete disregard for the business's long-term sustainability.

> *Private equity is like cholesterol: There is a good type and a bad type.*

The only strategic dimension of the deal is the investor's exit strategy. This approach was particularly prevalent in the easy-credit times before the global financial crisis of 2009, leading some A&D companies through three or four leveraged buyouts in a row, each time with substantial profits for the financiers.

Hedge funds' poor reputation is not completely undeserved, either. They tend to come and go opportunistically, based on their assessment of earnings multiples, "stock momentum" and "upside potential." Last year, TCI—one the largest hedge funds in the UK—wrote to Safran's CEO and CFO lambasting their management track record. The letter had the merit of bluntness: *"The Sagem merger was disastrous ... management's track record of investment ... is one of failure."*

But the letter also was judgmental: *"Bad decision-making... A true owner of the business would never have acted in such reckless manner"*, and of dubious veracity in some instances: *"Zodiac is a company of lower overall quality than Safran . . . the defense industry is in structural decline."* Even if there were some truth in the overall assessment—never mind that Safran's stock price has increased by 500% in the last five years—this type of letter, coming from an outsider with no long-term involvement in A&D, obviously does not generate warm feelings toward the financial community.

At the other end of the spectrum, though, a good private equity investor will offer a proper wealth preservation strategy for the original owners, boost investment and encourage ambitious growth plans, with a potentially transformative impact on the company and on the sector as a whole.

Many A&D suppliers clearly would benefit from opening up, at least partially, to such investors or, more generally, to financial markets. Both are a source of valuable capital. This is particularly important at a time when many subcontractors are undercapitalized and unable to make the large capital expenditures necessary to keep pace with surging commercial aerospace demand.

> *A good private equity investor will offer a proper wealth preservation strategy for the original owners, boost investment and encourage ambitious growth plans, with a potentially transformative impact on the company.*

Private equity can also help owners transfer the business smoothly to the next generation and bring new and vital skills to a company. Indeed, a lot of family-owned businesses are struggling to get to the next level simply because their owners do not want to relinquish any control. But sometimes, there is more value to be created for the founders and their descendants by taking a back seat and bringing in new shareholders.

The most inspiring example in Europe is probably Zodiac Aerospace, a world leader in aircraft equipment and systems including aircraft interiors. It was established more than 100 years ago as a family business but understood early on the value of finance to support its industrial growth strategy. Indeed, Zodiac was the first company to be listed on the "Second Market" of the Paris Stock Exchange in 1983, when its market cap was less than the equivalent of €50 million.

Today the company's market cap is more than €7 billion, but the founding families still own 24% of the shares and 35% of the voting rights. You can't argue with success.

Success Hard to Tell from Panic

April 2014

WITH A COMBINED BACKLOG of more than 10,000 commercial aircraft worth close to $1 trillion, Boeing and Airbus have never been in stronger positions, in particular when it comes to negotiating with their suppliers. They can pretty much promise them ten years of recurring orders on such programs as the Boeing 737 and 787 or Airbus A320 and A350. In exchange, the airframers expect significant investments, cost reductions, improved quality and close to 100% on-time delivery.

Only the largest suppliers have the ability to push back and manage their businesses on their own terms, if only to some extent. For the rest, it is more like "take it or leave it." In Europe in particular, the weight of Airbus in the aerospace sector is such that its loss as a customer is simply not an option.

> *In Europe, the weight of Airbus in the aerospace sector is such that its loss as a customer is simply not an option.*

This predominance is clearly a two-edged sword for Airbus. On one hand, it gives the company incredible leverage on its suppliers, as it can make or break almost any of them. No consolidation operation in the European supply chain today can happen without the blessing of Airbus, be it directly by acting as a "broker" of a deal or indirectly by bringing in some "friendly" investors, such as ACE Management, an investment firm that is majority funded by the Airbus Group and has been involved in most merger and acquisition deals in the French aerospace supply chain for the last few years.

But the downside of such dominance is that Airbus has no choice but to take responsibility for every one of its suppliers, however small. For historical reasons, OEMs like Airbus have been relying on very small companies as single sources for some key parts and sub-assemblies. When such small players are asked at the same time to invest in new capabilities and share the risks for a brand-new program like the A350, increase their production rates by up to 25% for the ongoing and booming A320 family, and take their businesses global to diversify their supplier and customer bases, they become overstretched both financially and operationally. Such companies can spin out of control quickly—and run out of cash—forcing Airbus to intervene in firefighting mode, even if it is only a $10 million company producing one small part for a single program.

Ideally, this job of managing the supply chain should be performed by the Tier 1 suppliers, Airbus' largest direct suppliers, but they are too vulnerable themselves to do it. Latécoère, for example, was identified six years ago as a potential consolidation platform in the European aerostructure sector. Airbus even considered selling the company two of its flagship factories that would have transformed it into a world leader, in particular with a breakthrough involvement in the A350 program. But then Latécoère overstretched itself and became so undercapitalized that it almost went bankrupt in 2010. Even today, its debt-to-capital ratio is 180% and the company has a huge €300 million debt that comes due next year. Nobody knows how the company will manage it.

> Beyond the glamorous and unprecedented backlogs of Airbus and the likes, the business reality down the supply chain displays a very different picture: too much fragmentation, too much debt, antiquated governance models, lack of international presence, and undercapitalization.

So beyond the glamorous and unprecedented backlogs of Airbus and the likes, the business reality down the supply chain actually displays a very different picture: too much fragmentation, too much debt, antiquated governance models, lack of international presence as well as undercapitalization, to list a few.

In a nutshell, too many weaknesses are embedded in the system that are not being addressed properly. This is simply because, as orders keep pouring in, OEMs have just one obsession: to keep the system running, faster and faster, like a steam engine being fueled with more coal than it can consume and therefore overheats dangerously.

Yet, in the middle of all this, some suppliers are riding high. Figeac Aero for example, a French aircraft parts company with €140 million in annual sales, is growing at 20% per year and generating a 20% EBITDA (earnings before interest, taxes, depreciation and amortization) margin, recently raised €15 million when it was listed on the Paris stock exchange. A few weeks later, the company was valued at close to €350 million and its founder and CEO became an instant star in the French aerospace community.

Never mind that the company's debt-to-capital ratio is 180% and that the founder still owns 94% of the company, therefore making the market cap rather meaningless. The story is appealing enough to make all other owner-CEOs in the sector believe they are sitting on piles of gold—and their bankers are more than happy to tell them just that. Rather than a pile of gold, it

Supply Chain Dynamics

looks more like a minefield to me. As the French painter Edgar Degas once said, *"There is a kind of success that is indistinguishable from panic."* That seems to describe quite well what is happening in the aerospace supply chain right now.

KEY CONCEPT

Supply Chain Structure

The commercial aerospace supply chain is a pyramid made up of several tiers. At the top are the aircraft **OEMs** (Airbus, Boeing, Embraer, Gulfstream…). Below them are hundreds of suppliers, some of them very large (Safran, UTC, GE…), many of them small, who sell their products directly to the OEMs and are therefore called **tier-one** suppliers.

OEMs have been aiming to significantly reduce the number of their tier-one suppliers, by raising the bar in terms of minimum size and ability to invest in new programs as 'risk-sharing partners'. This means that tier-one suppliers are expected to self-finance the development of their products in exchange for an exclusive supply contract for the lifetime of the program. These tier-one suppliers are in turn expected to manage their own supply chain made up of smaller players (**tier-two** and **tier-three** levels).

For the time being though, OEMs still rely on many small, highly dependent suppliers to directly procure thousands of parts and sub-assemblies, often as single sources (see Airbus example below), which is a factor of supply chain vulnerability, both in up cycles because many of these smaller suppliers struggle to keep up when volumes increase rapidly, and in down cycles, because these suppliers can quickly run out of cash, with a real risk of becoming a grain of sand in the gears of the OEM's entire production system. That is why OEMs often find themselves acting as firefighters (coming to the rescue of struggling suppliers), rather than as the architects (of an efficient and robust supply chain) they aspire to be.

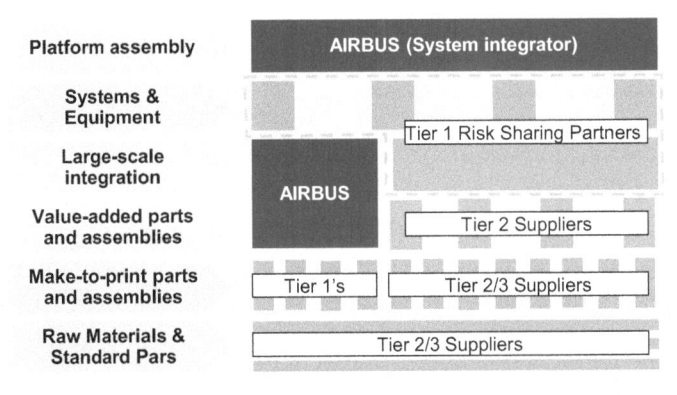

	AIRBUS (System integrator)
Platform assembly	
Systems & Equipment	Tier 1 Risk Sharing Partners
Large-scale integration	AIRBUS
Value-added parts and assemblies	Tier 2 Suppliers
Make-to-print parts and assemblies	Tier 1's · Tier 2/3 Suppliers
Raw Materials & Standard Pars	Tier 2/3 Suppliers

Source: Author

Chain Drive

April 2015

IN A RECENT AVIATION WEEK article, Airbus COO Tom Williams explains that while his company's backlog provides it a tremendous opportunity to negotiate better terms with suppliers and thus lower costs, he worries about the ability of suppliers—especially Tier 2 and 3—to keep pace operationally and financially with the rapid ramp-up to high-volume production on such models as the A350 and A320neo.

Indeed, there is a clear sense of panic in the European aerospace supply chain as Airbus keeps taking in new orders, raising its monthly production targets to unprecedented levels and asking suppliers for price reductions. While too many keep their heads in the sand and hope for the best, others are more proactive and engage in aggressive investment and performance improvement plans. The question, though, is not so much about who will die and who will thrive, but rather whether the overall system will survive. And this is where several misconceptions about supply chain competitiveness need to be brought to light.

The first of these misconceptions is that competitiveness of individual companies makes up the competitiveness of the whole supply chain, and therefore one should measure all suppliers' performances by the same benchmarks of reliability, productivity and cost efficiency to improve the overall performance of the system.

In complex systems such as the aerospace supply chain, each company is essentially a node in a large network of interconnected players.

This is not always true. In complex systems such as the aerospace supply chain, each company is essentially a node in a large network of interconnected players. What matters in the end is whether the network functions well as a whole. For that purpose, it is likely that companies will need to contribute differently to the efficiency of the network: some will excel at reliability, some at innovation, others at responsiveness or cost efficiency. Thus, applying the same formulas to all companies does not work. A small local supplier with limited capital and an expensive but highly experienced and responsive workforce, for example, may be as valuable as a large, multinational player with highly standardized but inflexible processes. Indeed, rapid prototyping and speed machining are increasingly important as OEMs continue tweaking designs throughout a program's lifetime.

The second misconception is that it is up to the suppliers to get to know their customers better in order to become more competitive. While customer intimacy is clearly very important, so is supplier intimacy. Essentially, companies, especially OEMs and large Tier 1's, should put as much effort into working with their suppliers as they do working with their customers. One of the foundations of the Japanese Keiretsu model's enduring success is the intimate knowledge of its suppliers that the OEM acquires by sending its engineers to spend time with them, at their sites, to better understand their business and way of working. This foundation is captured by the concept of "*genchi genbutsu*," which essentially implies looking at "actual workplaces and actual things." I doubt many aerospace OEMs achieve that level of engagement with their suppliers, and that is why there is still a high level of mistrust and "win-lose" mentality—hence inefficiencies—embedded in the supply chain.

The third misconception is that competitiveness is all about processes. Lean manufacturing, six-sigma, statistical process control, etc. are all important operational processes, but they are increasingly standard and therefore undifferentiating. They are just qualifiers. The reality is that in complex, tightly linked systems such as the aerospace supply chain, problems and accidents are inevitable, regardless of how good and under control the processes are. The difficulty is knowing how to react when such problems occur. And that comes down to people management skills and, ultimately, human interactions: trust, respect, honest communication (not electronic but actual face-to-face conversations) and the ability to listen, display empathy and share information.

To simplify, one could say that while processes are about doing things right, human relations are about doing the right thing, which is the ultimate differentiator between failure and success, especially when systems or processes start breaking down.

> *While it is easy to single out individual companies as risk factors for the greater supply chain's competitiveness, the real challenge is at the level of the overall ecosystem that encompasses all players, from top to bottom.*

While it is easy to single out individual companies as risk factors for the greater supply chain's competitiveness, the real challenge is at the level of the overall ecosystem that encompasses all players, from top to bottom. The main jeopardy for the aerospace supply chain is primarily a system risk, and its long-term competitiveness will be driven by the ability of the big players— OEMs and large Tier 1 companies—to question their assumptions about what makes the supply chain competitive and rethink the way they allocate their resources accordingly. Success will not come from fighting tens of small individual battles, supplier by supplier, but from winning one big hundred supplier war.

KEY CONCEPT

GENCHI GENBUTSU

Genchi genbutsu is a Japanese concept meaning "Go and See for Yourself" or "Going to the Source". It was coined as part of the development of the **Toyota Production System** by Taiichi Ohno.

The purpose of *genchi genbutsu* is to illustrate the importance of being on the factory floor where production is happening so that you can see both the creation of value and the generating of waste.

Ohno used to emphasise this point to new hires by taking them down to the factory and drawing a chalk circle on the shop floor. The young engineer would then stand in this circle and be told to observe the world around them, then report to him everything they saw. Those who did not learn enough would have to repeat the process.

Genchi genbutsu thus describes the practice of finding your answers right down at the source, rather than relying on second-hand reports or charts of data to achieve true understanding. This practice emphasizes going to a place (*gemba*) where work is actually done and where you watch, observe, and ask questions.

REVISITING THE CUSTOMER-SUPPLIER RELATIONSHIP MODEL

DECEMBER 2015

IN SPITE OF MULTIPLE INITIATIVES to improve the efficiency of the supply chain, current models of the customer-supplier relationship are not sustainable.

As aircraft OEMs keep changing their production forecasts while raising their on-time delivery requirements—Airbus recently announced changes in the A320 conventional engine option/new engine option mix and an A380 production rate reduction—suppliers are becoming increasingly frustrated with the level of support and communication they receive from their main customers. Similarly, in the defense sector, a lot of suppliers do not understand why their customers keep asking for price reductions without providing more visibility into their long-term plans.

In spite of prime contractors implementing diverse tools that are meant to improve relationships with suppliers (such as digital cooperation platforms, quality improvement programs or field engineering support), recent evaluations of customer-supplier relationships highlight a huge gap between the "talk" at the strategic level and the "walk" at the working level.

> *Recent evaluations of customer-supplier relationships highlight a huge gap between the "talk" at the strategic level and the "walk" at the working level.*

Key complaints on both sides include a lack of anticipation, poor resource management, too much bureaucracy on the customer side and lack of agility on the supplier side, leading to a still pervasive "win-lose" type of relationship.

I believe the reason why prime contractors are so often at odds with their suppliers comes down to an outdated approach to problem-solving. Let me explain why.

Most companies today still function in a linear, basic "think-act" mode. First, they think through a problem and try to generate ideas to solve it, then they move on to implementing solutions that come out of their idea-generation work. This type of approach usually brings disappointment, and its effects fade away rapidly because: 1) the people involved in the "acting" part are typically not the same as those involved in the "thinking" part, hence much is wasted in the transfer from one to the other; and 2) this linear process does not match the way an organization actually works, which is much more iterative and cyclical, if not erratic.

In order to transform an organization or, in the case of the aerospace supply chain, an ecosystem, one first needs to work on two fundamental dimensions: the vision and the belief; the vision of the end game one wants to reach, and the belief that it is achievable. Let's take the concrete example recently encountered of a naval defense prime contractor that is faced with a near-monopolistic supply base and wishes to develop a new supplier relationship model to gain more leverage and control of its overall cost base. Such a prime will typically assume this situation is a major constraint and feel it is being held hostage to its suppliers. Accordingly, it will design a model aimed at minimizing such constraint. When it begins to implement the model, though, it is likely to find itself in a conflictual relationship with suppliers that feel threatened by the new model, often leading to a breakdown in communication and major tensions. This is the frequent outcome of the traditional "think-act" problem-solving mode.

To transform an organization or, in the case of the aerospace supply chain, an ecosystem, one first needs to work on two fundamental dimensions: the vision and the belief.

However, in the alternative mode—let's call it "see-believe-think-act"—the prime contractor will first work, possibly in cooperation with its suppliers, to envision a relationship in which the same near-monopolistic supply base actually works to its advantage and is an asset—in particular due to increased mutual trust and understanding, better communication and information transparency.

Once such a "win-win" scenario can be visualized and played out (at least in people's heads), then the parties start changing their beliefs about what is possible. What had been seen as a constraint by the customer and as a way to maximize profits (to the detriment of the customer) by the supplier becomes an opportunity to improve long-term performance for both. Thereafter, the idea-generation work will be about optimizing an opportunity, rather than minimizing a constraint, and the transition to action will be cooperative rather than conflictual, hence more sustainable.

Ultimately, the only way to make the customer-supplier relationship sustainable is to develop a model based on mutual trust and long-term commitment. In that respect, the Japanese Keiretsu model, in its digital "2.0" version, far from being outdated, is probably the most relevant model because it relies on solid principles of supplier intimacy (on the customer's part), continuous performance improvement, healthy competition among suppliers and smart information-sharing. Aerospace and defense OEMs will not have a fully secured supply chain until they purposely and decisively embrace such an approach, based not only on "thinking and acting" but foremost on "seeing and believing."

Chapter 3

(GEO)POLITICS

"We live as a nation under an enormous danger, for we are at the mercy of foreign nations, not only for our food and for our clothes, but also for our huge invested capital."

E.E. Williams, British journalist (1866–1935)

CHAPTER INTRODUCTION

The A&D industry is highly political and strategic in many countries, starting with those with the most at stake, namely the US, Western Europe (more specifically the U.K, France and Germany), China and Russia.

This chapter illustrates this geopolitical dimension by putting in contrast the ambitions of a new aerospace power – China – with the struggles of an old one – The United Kingdom. It also contrasts the pragmatism of Germany's export policy with the failings of France's political leadership.

Overall, it shows how political decisions impact each country's industry success and, specifically for the UK, how the ongoing exit from the European Union (EU) has created a new challenge for both the British industry and the EU as political dreams clash with economic reality.

CHINA'S AEROSPACE DREAM EDGES CLOSER TO REALITY

SEPTEMBER 2017

CHINA'S COMMERCIAL AEROSPACE ambition has long been known, but a few recent milestones are bringing it closer to realization. Since the establishment of AVIC 1 and AVIC 2 in 1999, which really marked the beginning of China's push into the commercial aerospace market, there has been no doubt about China's intent and the direction it wants its industry to take. But the pace at which it would achieve this ambition was less certain. Almost 20 years on, things have become much clearer, and there is now little doubt that China is well on its way to succeeding where several other countries have failed, in becoming a full-fledged player in the large commercial aircraft manufacturing sector, with the potential to disrupt the long-standing duopoly of Airbus and Boeing.

> *There is now little doubt that China is well on its way to succeeding where several other countries have failed, in becoming a full-fledged player in the large commercial aircraft manufacturing sector.*

To be fair though, no other country has ever had China's assets: a stable government with an unequivocal, long-term financial and strategic commitment and a huge domestic—hence mostly captive—market. According to market forecasts, China's domestic air traffic is expected to almost quadruple between now and 2036 to reach 1.6 billion passengers, which will be more than twice the US's domestic traffic by then. And while there is still a long way to go before airplanes designed and made in China compete on the world stage with the likes of the Boeing 737 or Airbus A320, the momentum is there, and the country's aerospace industry is entering a new phase of development and maturity.

Most observers would think the single-aisle COMAC C919's first flight last May is illustrative of this, but while it is a key milestone, it does not represent the whole story about what is happening in the industry. Two other events are much more telling.

The first is the establishment of the Aero Engine Corp. of China (AECC) last year, which consolidates Chinese aero-engine know-how and capabilities, indicating that the indigenous development of aircraft engines is now a strategic priority for China. This focus is even more striking when one realizes that China now has access to rhenium, a rare metal that helps create strong superalloys necessary for the manufacture of high-pressure jet engine turbine blades. Since the 2010 discovery in the Shaanxi province of reserves, which account for 7% of the world's total, considerable effort has been made to exploit and process this metal. Moreover, a major technological milestone was recently achieved that opens the door for mass production of single-crystal turbine blades, a key component of modern jet engines. It is thus no surprise that one of the cornerstones of the newly established joint Sino-Russian widebody aircraft program is the development of a dedicated, state-of-the-art engine.

The second event of note is COMAC's latest round of financing—it raised 15 billion yuan ($2.3 billion) last month in the form of a 10-year debt investment plan—combined with the memorandum of understanding (MOU) signed in June by Airbus and the National Development and Reform Commission (NDRC). The financing and MOU are intended to help bring about a fully developed, competitive domestic supply chain, the former through the injection of research and development money down the supply chain and the latter through the integration of Chinese suppliers in Airbus's global supply network. The objective, as outlined in the "Made in China 2025" plan, is for Chinese suppliers to provide 80% of all parts by 2025.

Taken together, these events and strategic initiatives highlight the Chinese aerospace industry's realization that, in order to become a globally competitive player, it will need depth as much as breadth. After getting their hands on both ends of the value chain—aircraft design and final assembly—the Chinese now understand that what will make or break their industry over the long term is what happens in the middle of the value chain, at the component and subsystem levels. And that is why China's ambition to compete with Boeing and Airbus, as well as with GE and Rolls-Royce for aircraft engines, is now more credible than ever.

> *After getting their hands on both ends of the value chain—aircraft design and final assembly—the Chinese now understand that what will make or break their industry over the long term is what happens in the middle of the value chain.*

By investing deep in the industrial fabric of the country, from the coastal cities of Tianjin and Shanghai to the new industrial frontiers of the Western provinces of Shaanxi and Sichuan, from the processing of rare metals to the manufacturing and assembly of complex aircraft parts, China is adding substance to form and showing the world that, as far as its commercial aerospace achievements are concerned, the best is yet to come.

A Proud Past but Uncertain Future (UK Aerospace)

June 2016

IN LESS THAN A MONTH, British, Irish and Commonwealth citizens living in the UK will vote to decide whether the UK should remain in the European Union or not. While the stakes are high, this is really business as usual for the British, as this June 23 "Brexit" referendum is just the latest episode in the country's longstanding ambivalence toward Europe.

In many ways, the recent history of the UK aerospace sector reflects such uncertainty: After being involved in the very early stages of what became Airbus in the 1960s, the British government withdrew from the project in 1969, leaving Germany and France to proceed on their own. Ten years later, in 1979, it returned—via state-owned British Aerospace—by acquiring a 20% share of Airbus. Twenty years later, in 1999, British Aerospace passed up an almost sealed deal with DaimlerChrysler's aerospace division, which would have created Europe's largest aerospace conglomerate, and instead merged with fellow British company GEC-Marconi, which itself eschewed a potential tie-up with France's Thomson-CSF (now Thales). From that point on, the newly formed BAE Systems directed most of its resources toward penetrating the US defense market.

This US-centric strategy, shared by most UK A&D companies at the time, culminated in the sale of BAE's Airbus stake in 2006. This turned out to be a strategic mistake, and in 2012 BAE was willing to pay a significant premium to buy its way back into Airbus through a merger with EADS, which ultimately fell through.

Partly because of all these mistakes by the firm that was formerly regarded as the national champion, the UK aerospace industry is in a much weaker strategic position today than 30 years ago. Essentially, it has lost its "head"— that is, an organization capable of designing a complete aircraft. This capability is gone, and BAE Systems is the shadow of its former self in terms of industry leadership and aerospace design engineering capabilities.

While British Aerospace (now BAE Systems) was historically the "head," Rolls-Royce has always been the "heart" of the country's aerospace sector.

While British Aerospace (now BAE Systems) was historically the "head," Rolls-Royce has always been the "heart" of the country's aerospace sector. Going back to the company's early days, when talented engineer Henry

Royce and flamboyant race car driver Charles Rolls joined forces to create what would quickly become an iconic brand, the company's mystique took root in the national psyche. Royce in particular was described by a journalist as an "engineering artist" who "cares more about getting the best out of an article than he does about getting the most money out of it."

During World War II, as the firm channeled all its energy into the war effort—*"Work until it hurts!"* exhorted the company's chief executive to its workers—it became the symbol of British resilience in the face of adversity. Thus, Rolls-Royce's current troubles feel like the British aerospace industry being wounded in its heart.

Meanwhile, the part of the industry that does work well is the one that builds aircraft parts but does not design them: Tens of small- and medium-size manufacturers of "build-to-print" aerostructure and detail parts are today the industrial backbone of UK aerospace. But very few of them, if any, design their own products.

As an illustration, only one British company today—GKN—is qualified by Airbus as a tier-1 supplier of "design and build" aerostructures. Without strong domestic design and system engineering capabilities, the industry must compete with an increasing number of nations for the award of work packages and for foreign investment. And while the British have always displayed a smug optimism about their ability to attract foreign investors, such reliance is as dangerous today as it was 100 years ago when a popular commentator, E.E. Williams, said that *"we live as a nation under an enormous danger, for we are at the mercy of foreign nations, not only for our food and for our clothes, but also for our huge invested capital."*

Like the country as a whole, the British aerospace industry is at a crossroads: in search of its true identity, torn between the pragmatic recognition that its future is inextricably tied to the future of Europe as a whole and the deep-rooted belief that it is destined for greater achievements on the world stage, for which its European membership may be perceived as a drawback.

Having already been deprived of its head, now struggling with its heart, let us just hope that the British aerospace industry will not end up losing its soul.

Merkel's Mark

February 2014

TWO RECENT EVENTS EPITOMIZE radically diverging fates in the global defense industry. At the end of 2013, Singapore ordered Type 218SG submarines from German company ThyssenKrupp Marine Systems (TKMS) for €1.6 billion ($2.2 billion). A month later, the French defense minister awarded a €1 billion contract for the modernization of Dassault's Rafale fighter aircraft. A good story for both companies? Not quite.

The reality is that TKMS's Singapore order is a genuine success story, while the Rafale contract is a rather sad face-saving exercise. Some observers would say it is merely a reflection of German efficiency versus French pride, and they might not be completely wrong. But more than that, it is a reflection of German pragmatic diplomacy versus French political incompetence.

The last few years have indeed been particularly good for the German defense industry. The sale of submarines to Singapore is only one of many export deals recently signed by German contractors in Asia and the Middle East. Among the best-sellers are Krauss Maffei Wegmann's Leopard 2 tanks and Boxer vehicles and Heckler & Koch's assault rifles and machine guns. In 2011, Germany became the largest European exporter of defense equipment, ahead of France and the UK Not bad for a country that has prided itself on its restrictive defense export policy.

In reality though, the government of German Chancellor Angela Merkel has been as business friendly as possible. Under a new stance that supports military exports to "partner countries" rather than sending German soldiers to war or unstable regions, it offers government export guarantees without too much arm twisting, This so-called "Merkel doctrine" also represents a clear effort to help the domestic defense industry at a time when German armed forces are drastically downsizing.

> The "Merkel doctrine" represents a clear effort to help the domestic defense industry at a time when German armed forces are drastically downsizing.

Interestingly, the biggest beneficiaries are groups such as Rheinmetall, Krauss Maffei Wegmann, Diehl or TKMS, which have remained very German in their culture and ownership. This contrasts with EADS which, in spite of strong German roots, is not considered a German company anymore and therefore does not get as much support from the government. It is the same in France, where EADS finds itself at a disadvantage when it comes to obtaining government support compared to "more" French companies such as Dassault Aviation.

While Dassault's Mirage 2000 aircraft was a great export success thanks to its versatility and performance, Rafale is a story of stubbornness and manipulation. Twenty-eight years after company founder Marcel Dassault—then 93 years' old and disclosing the new aircraft to the public for the first time—claimed loud and clear that the Rafale would be a "world aircraft." It has not won a single export contract.

Rafale's combat performances have been proven in Afghanistan, Libya and Mali, but it comes across as over-engineered, with a price tag that makes it inaccessible for most countries. The recent €1 billion modernization contract is proof of its commercial failure. It is a last-ditch attempt to give Rafale a boost, hoping a proposed sale to India will materialize soon and orders from Middle Eastern countries will follow.

German companies understood long ago that winning in the export markets was not about offering the best technology but just the right product at the right price and at the right time. But French companies have had great export successes as well. The problem is that France has been without strong political leadership or any defense-industrial strategy for decades. In that respect, the French government is like a puppet, disjointed, inconsistent and easily manipulated by interest groups that ultimately neutralize each other.

There is not much French defense companies can do about the incompetence of their country's political leadership. But one thing is clear: They, like their counterparts in other European countries, must take their destiny into their own hands. They cannot wait for their government to rescue them.

German companies understood long ago that winning in the export markets was not about offering the best technology but just the right product at the right price and at the right time.

These companies have to go where the market is, offer competitive products and services, and hope their government will support them when needed. They must go as far as designing products for the export markets first and then use the proceeds of export sales to fund better systems for their national customers. That will be a true cultural revolution for many, but one surely worth undergoing.

British A&D in Shambles

March 2018

IN JUNE 2016, ON THE EVE OF the Brexit referendum, I wrote that the British aerospace industry was at a crossroads, faced with the prospect of a political outcome that could seriously compromise its future. Almost two years later, one can now officially say that the British aerospace and defense sector (A&D) altogether, including the British Armed Forces, is in shambles.

Not a week goes by without some senior industry executive, highly ranked military official or politician commenting on the dangers of a bad Brexit deal, the worrisome state of the defense forces or the frailty of the British industrial base. Of course, not all of that is directly related to Brexit, but the process of exiting the European Union is destabilizing an industry that was already struggling (*see graph below*) and exacerbating budgetary uncertainties that, coupled with weak political leadership, contribute to the general malaise.

> *The process of exiting the EU is destabilizing an industry that was already struggling and exacerbating budgetary uncertainties that, coupled with weak political leadership, contribute to the general malaise.*

For a start, the Armed Forces are trying to make sense of the government's confusing signals and to articulate a coherent strategy for a badly needed modernization. After 15 years of continuous presence in Iraq, Afghanistan and more recently Syria, the forces are beset with a serious "war fatigue" syndrome; their operational equipment is aging (such as Tornado fighter jets, Duke frigates, Warrior vehicles), and recent investments in new capabilities (the F-35, aircraft carrier, Ajax vehicle, cybersecurity, and such) have not yet borne fruit. Gen. (ret.) Richard Barrons, former Commander Joint Forces Command, bluntly stated that the British Armed Forces were close to breaking, with the Royal Navy underfunded, the Royal Air Force at the edge of its engineering capacity and the British Army 20 years out of date.

Defense Secretary Gavin Williamson has tried to reassure the defense community by announcing that the UK will soon launch a "Combat Air Strategy" aimed at maintaining a world-leading combat air capability, but the statement came across as hollow and noncommittal when the French and Germans have signaled their intention to jointly develop a future "European" fighter jet.

(Geo)Politics

Meanwhile, clouds continue to darken for British defense companies. Their historical leader, BAE Systems, has been in a strategic dead end for years, overdependent on unpredictable and sparse government contracts and looking powerless in the face of a changing environment, as if it had never recovered from the failed merger with EADS. Other major defense players also are embroiled in existential crises—Cobham, Babcock, Chemring, even Ultra Electronics, one of the historical stars of the sector, whose long time CEO had to resign following several years of disappointing results.

On the commercial aerospace front, things are not any brighter, although the tentative transition deal gives industry until 2020 to reach a trade agreement. All stakeholders agree that Brexit will have damaging effects on the industry's global competitiveness. Airbus is increasingly worried about potential supply chain disruptions once the UK is officially out of the EU after March 2019. With 15,000 employees across 25 sites and £5 billion ($7 billion) spent every year with UK suppliers, the stakes are huge. This spurred Klaus Richter, Airbus chief procurement officer, to request that all suppliers establish contingency plans to ensure deliveries will not be impacted.

All this would still appear manageable if it did not come at a time of great stress on the British supply chain. Rolls-Royce, the country's biggest commercial aerospace supplier, is struggling to fix its Trent 1000 engine, and GKN, the second largest, is fending off a hostile takeover bid from corporate

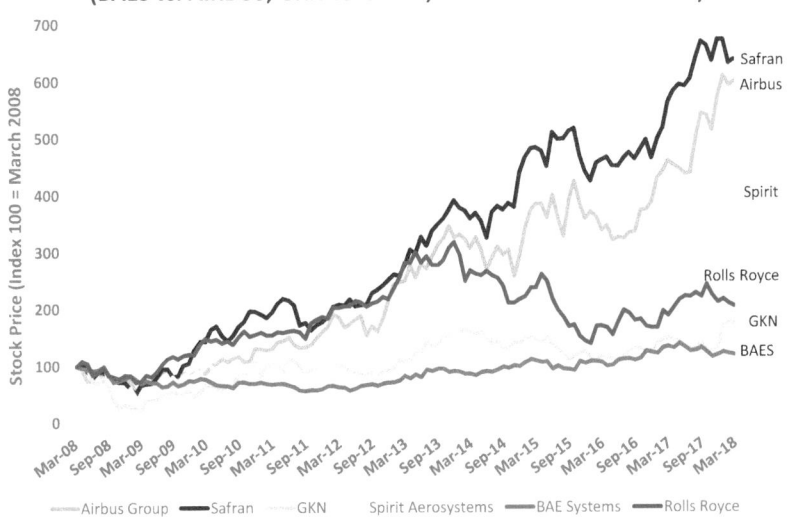

2008-2018 STOCK PRICE PERFORMANCE OF BRITISH A&D PLAYERS vs. PEERS (BAES vs. AIRBUS, GKN vs. SPIRIT, ROLLS ROYCE vs. SAFRAN)

Source: Yahoo! Finance

raider Melrose plc. GKN's perceived potential breakup prompted a group of MPs (Members of Parliament) to call on the government to block the deal, citing national security and arguing the takeover would mean the end of one of the UK's industrial jewels.

It is ironic that many in the British business community see GKN as a last bastion of UK engineering excellence. Ten years ago, no one would have blinked an eye at the prospect of it being acquired, let alone by another UK-based company.

Sadly, it is a reflection of the anxiety that surrounds the British aerospace and defense sector today. As politicians march blindly toward an ominous end to the Brexit process, business stories such as GKN's or Airbus' remind us that, as much as British people cherish their independence, the only pragmatic and constructive way forward for their aerospace and defense players is "in," not "out."

Galileo's High Stakes

June 2018

AMONG THE MANY DISPUTES that have embroiled the UK and the European Union regarding Brexit, one centers on the participation of the UK in the Galileo program, the $12 billion European Global Navigation Satellite System (GNSS) being rolled out to complement and possibly supersede the US GPS system.

The UK, which has been involved in the project since its onset by contributing payloads, the ground control segment and security software, wants to stay in the program, while the EU argues that, once out of the EU, the UK should have reduced status and access, in particular to the classified part of Galileo called the PRS (Public Regulated Service), which is critical to the security of the European member states.

In response, the UK government officially announced May 23 that, while it would still prefer to continue participating in Galileo, it is "exploring alternatives to fulfil its needs for secure and resilient position, navigation and timing information, including the option for a domestic satellite system."

What should we make of this dispute?

First, it is mind-blowing that a country such as the UK can make as drastic a political decision as leaving the EU and hope that there would be no economic, industrial, or financial consequences. In this particular instance, we are talking about at least $5-7 billion to design and build a new GNSS constellation. Can the country afford it? Barely. Would it be public money well spent? Hardly, considering the huge budgetary needs in other areas of public policy such as health care, education and housing.

> *It is mind-blowing that a country such as the UK can make as drastic a political decision as leaving the EU and hope that there would be no economic, industrial, or financial consequences.*

Besides, it is likely that an agreement can be struck with the EU along the lines of those already in place with Norway and Switzerland, two countries that are not member states, but which have access to Galileo services, including its secured PRS part.

For the UK, however, this type of user agreement would not be considered enough, mainly because the country's space industry has been involved in developing and building some sensitive parts of the system and the UK believes this warrants a special status in the program, even after it exits the EU.

This seems to be a relatively weak argument if one recalls that the most critical part of the Galileo satellites—the atomic clocks—are designed and made in Switzerland, a non-EU country.

So the UK really has no one but itself to blame for the situation, and it would seem logical that its future status will be downgraded. Having said that, the dispute also highlights the strategic nature of Galileo, and more generally of the GNSS as the main and most accurate source of position, navigation and timing (PNT) information in the world today.

A report published by the UK Government Office for Science in January made clear the country's high level of dependency on the GNSS not only for critical national infrastructure such as computer networks, emergency services, communications, electricity transmission and financial transactions, but also for future digital infrastructure such as 5G, autonomous vehicles and the Internet of Things.

Having reliable access to the GNSS has become a matter of national security, all the more as the GNSS-derived signals are typically vulnerable to interference through "jamming" and "spoofing"—which was once the domain of states but is now within the capability of hackers, pirates or terrorists. This has created the need for a higher level of GNSS performance, dubbed "Resilient-PNT."

What this means is that GNSS constellations are now a strategic weapon for the countries that own them—namely, China, Russia, the US and the EU—and a matter of survival for all the critical infrastructure and service operators, including military forces, that use them. In this context, it is understandable that the UK is genuinely concerned about its future status in Galileo.

The Galileo dispute is a unique opportunity for the EU to develop a new blueprint for security, defense and space cooperation with third countries, of which the UK could be the paragon.

Overall, though, the Galileo dispute is a unique opportunity for the EU to develop a new blueprint for security, defense and space cooperation with third countries, of which the UK could be the paragon. Behind the rhetoric, it is in the mutual interest of the UK and the EU for the British to stay involved in the program. Therefore, it would make sense to use Galileo as a springboard to design a new partnership model whereby the European Union, while reinforcing its internal cohesiveness through a more integrated defense and space policy, allows for more flexibility when dealing with third-party countries like the UK.

Increased intra-European integration and enhanced extra-European modularity should go hand in hand and are the only viable "win-win" outcome to an otherwise calamitous political situation.

KEY CONCEPT

Partnership Model

Insofar as the status quo is not satisfactory to anyone, Brexit could play the role of a catalyst for the development of a new model of international defense partnership, more in line with the geopolitical realities of the 21st century. Such model would aim to bring into coherence the objectives and resources of all stakeholders (individual countries and international organizations) in a relatively flexible manner while maximizing opportunities for cooperation.

However, in the case of the UK and the EU, two conditions need to be met for such a "win-win" scenario to materialize. On the one hand, the dynamics of "**intra-European integration**" must be strengthened (from this point of view the establishment by 25 EU member states of a Permanent Structure Cooperation -PESCO- goes in the right direction). On the other hand, "**extra-European modularity**" – with NATO, the United Nations and other third parties – must also improve significantly, for example through more interoperability between forces and the development of bilateral and regional cooperation programs, whether institutional or ad hoc (such as the Franco-British treaty of Lancaster House, the "Joint Expeditionary Force", the "European Air Transport Command "or NATO's "Framework Nations Concept"). All other scenarios will likely lead to a weakened position for the UK or the EU, or for both.

POST-BREXIT SCENARIOS FOR UK-EU DEFENSE RELATIONSHIP

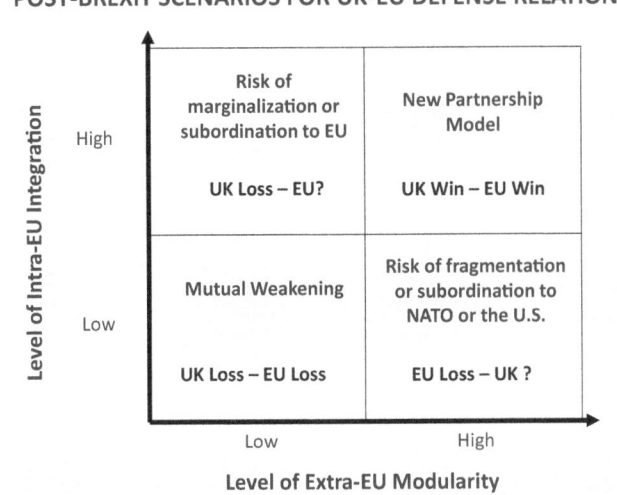

Level of Intra-EU Integration	Low Extra-EU Modularity	High Extra-EU Modularity
High	Risk of marginalization or subordination to EU — UK Loss – EU?	New Partnership Model — UK Win – EU Win
Low	Mutual Weakening — UK Loss – EU Loss	Risk of fragmentation or subordination to NATO or the U.S. — EU Loss – UK ?

Source: Author

Part Two

CORPORATE STRATEGIES

Chapter 4

BUSINESS FUNDAMENTALS

"For individuals, character is destiny. For companies, culture is destiny."

Tony Hsieh, American entrepreneur and investor (1973–2020)

CHAPTER INTRODUCTION

This chapter addresses four key strategic themes that constitute fundamental drivers of competitiveness, in general and specifically for A&D companies.

Corporate culture: Corporate culture is probably the ultimate differentiator when it comes to long-term business performance. Two companies with outstanding track records – Rockwell Collins and Zodiac Aerospace – illustrate how a strong and authentic culture can go a long way toward helping an organization achieve sustainable long-term excellence.

Geographic Clusters: As the aerospace market has become more global, it has become increasingly vital for companies to strengthen their home base and leverage the benefits of deep-rooted location-specific linkages. More than ever, local or regional clusters are a fundamental basis of competitiveness in the aerospace industry, both for small suppliers and large OEMs.

Commercial craftsmanship (The "art and craft" of selling aerospace products): A great technology does not always guarantee commercial success. But when a product does thrive commercially, its success is usually attributed to innovation and technical expertise. Often overlooked are the critical roles played by sales executives, whose craftsmanship, personality, and ability to connect with a customer at a personal level can still make a difference, even in technology- or data-intensive market environments.

Managing the political risk: The A&D business does not happen in a political vacuum. In fact, when it comes to major international contracts, politics often trump economics. It is therefore important for companies to incorporate the political factor into their overall business strategy. And the best way to develop a strategy resilient to the unpredictability of international politics is to create business interdependencies that will make unilateral political decisions more difficult to implement.

CULTURE IS DESTINY

JUNE 2013

EVERY YEAR WHEN AVIATION WEEK releases its Top-Performing Companies rankings as it did last month, the few of us on the Aviation Week's TPC Council of Advisers are never short of explanations about why a certain company comes out on top: operational excellence, smart business portfolio management, "sweet-spot" positioning in the value chain, disciplined financial management, etc. All these factors may indeed translate into superior annual performance metrics such as return on invested capital, return on sales or asset turnover. However, we often fail to talk about one of the most important competitive assets of a company: its culture.

The omission is easy to understand. There is no financial indicator to measure the strength of a corporate culture, yet it is probably the ultimate differentiator when it comes to long-term business performance.

> *There is no financial indicator to measure the strength of a corporate culture, yet it is probably the ultimate differentiator when it comes to long-term business performance.*

In that respect, it is worth stepping back from the circus of quarterly and even annual results to reflect on what makes a strong company culture and why it often translates into superior business performance. Two companies—one American and one European—epitomize what I mean by strong corporate culture. The first is Rockwell Collins. Even though by itself it is still a young company, having been established on its own in 2001 through a spinoff from Rockwell International, its roots go back to the founding of the Collins Radio Co. in Cedar Rapids, Iowa, in 1933, during the Great Depression. Beyond its excellent products and technologies, Rockwell Collins stands out for the way it has consistently remained strategically focused and faithful to its core values, in spite of major external events that easily could have derailed its early development.

Those values include commitment to research and development (reinvesting 18–20% of revenues every year) and bolt-on acquisitions, which have been two key strategic drivers. But the one that probably has had the most impact has been a relentless "focus on customers," which as former CEO Clay Jones puts it, "include the people who pay the money for our products, our shareholders, our community and our employees." Attached to that broad customer focus are a diversity and inclusion program as well as an overall engagement strategy that brought about such initiatives as a policy to encourage hiring of young veterans and far-reaching mentoring programs.

One could not talk about Rockwell Collins's corporate culture without underlining the importance of the company's leadership and commitment to those values. The fact that Jones, who retired last year, had been with the company for 20 years before running it as CEO for 15 more years surely goes a long way toward explaining why those values are so deeply embedded. Another important factor has been the careful external growth strategy that has led Rockwell Collins to favor small bolt-on acquisitions to large mergers. This has clearly facilitated the integration of newly acquired companies without damaging or diluting the existing culture.

On the other side of the Atlantic, a relatively similar story developed at French aerospace supplier Zodiac Aerospace. Even if the origin of the company dates back more than 100 years, it really started in 1973 when Jean-Louis Gerondeau came in as a young CEO. At the time, the company had revenues of the equivalent of €8 million and 400 employees. When he retired in 2008, its revenues then were more than €2 billion and it had 15,000 employees (its size has doubled since then).

Gerondeau nurtured a strong entrepreneurial culture that aimed at preserving the spirit of a small family business while becoming a world leader in aircraft equipment through the acquisition of tens of companies, none of them big enough to destabilize Zodiac's culture. Together with values such as humility, realism and respect ("for what has been accomplished, what is underway and what remains to be achieved"), this entrepreneurial spirit has contributed substantially to the company's consistent financial and operational performance and sets it apart from its competitors.

In 2010, after fellow French group Safran attempted to acquire the company, Zodiac was quick to reject the offer. Its entrepreneurial culture was incompatible with Safran's state-owned mentality. The rejection highlighted the risk of significant value destruction from combining two companies with different, if not opposite, cultures. That is what happened when Safran was created by the 2005 merger of Snecma with Sagem.

A strong and authentic culture goes a long way toward helping an organization achieve sustainable long-term excellence.

Indeed, while a strong and authentic culture goes a long way toward helping an organization achieve sustainable long-term excellence, a dysfunctional or incompatible culture can severely undermine business performance, especially when times get hard and resilience is needed. As Zappos.com's CEO Tony Hsieh puts it: "*For individuals, character is destiny. For companies, culture is destiny.*"

Clusters of Vitality

July 2013

WALKING THROUGH THE PARIS AIR SHOW last month, I was struck by the large number of companies exhibiting under one regional or cluster umbrella. There were, to mention a few, Aerospace Valley, Rockford Area Aerospace Network, Monterrey Aerocluster Mexico, Isle of Man Aerospace Cluster, Aero Montreal, Skywin Wallonie and Northwest Aerospace Alliance. Such groupings have a basic economic rationale: They allow small suppliers to be present at a show without bearing the full costs of renting their own stands in an exhibit hall. But there also is a deeper meaning to this profusion of the clusters.

An aerospace company is about more than machinery, hangars, engineering software or a distribution network. It is also defined by culture, tradition and relationships that have been passed through several generations of workers.

As the aerospace market has become more global and companies are stretched in many directions—sourcing here, outsourcing there, selling everywhere—it has become increasingly vital for them to strengthen their home base, as a way to preserve their true identity and uniqueness. Indeed, the skills and know-how involved in aerospace design and manufacturing are deep-rooted in the places where they were born and nurtured over several decades, if not a century. An aerospace company is about more than machinery, hangars, engineering software or a distribution network. It is also defined by culture, tradition and relationships that have been passed through several generations of workers, most of the time in one specific location, be it Seattle, Toulouse, Cincinnati, Filton, Wichita, Derby, Hamburg, Montreal, Fort Worth, St Louis, Long Beach, Ottobrun or East Hartford. Each of these locations is synonymous with one major aerospace player and, by extension, with a long history of aerospace know-how and a deep network of local partners.

This is why it is so important for original equipment manufacturers (OEMs) to keep their suppliers close while at the same time encouraging them to extend their footprint internationally. Airbus, Boeing, and others have understood that even in a global supply chain, there is value in geographical and cultural proximity. Being close makes it easier to communicate, identify and address problems swiftly and to get the responsiveness required for such critical tasks as rapid prototyping or speed shopping.

Additionally, thanks to the cluster effect, skilled labor is readily accessible, and there is a critical mass of experience and know-how ready to be employed. For the suppliers, sustaining a strong local presence close to their main customers while becoming more international is a balancing act that is hard to achieve—but critical to their success.

More than ever, local or regional clusters are the fundamental basis of competitiveness in the aerospace industry, both for small suppliers and large OEMs. Industry leaders in Europe and North America should not forget that. It is fine to go to China, India, Brazil or Russia to become a global player, but one should not underestimate the importance of strong local roots in shaping a company's future.

Companies can be acquired, merge with others and change names, but behind each of them there will always be one specific site or location that defines it more than anything else: Seattle for Boeing, Toulouse for Airbus, Montreal for Bombardier, Derby for Rolls-Royce, and Cincinnati for GE Aviation.

Some observers might say these deep roots act as "barriers to exit", tying companies down and negatively impacting their ability to reinvent themselves to adapt to shifts in the market. But I believe the "barriers to entry" they represent for potential entrants are much higher. It is one thing to build a new assembly line in Asia. It is another to replicate the intricacies of a network of relationships and partnerships that has been nurtured over several decades.

That is the strength of clusters: They cannot be easily replicated, and they take time to build, because the strength is not only about geographical location; it is also about relationships, cultural fit, and strategic intent.

> *That is the strength of clusters: They cannot be easily replicated, and they take time to build.*

Belonging to a cluster should be an essential part of any aerospace player's strategy, be it around an OEM or a large Tier I or smaller supplier. Not only will a cluster allow companies to aim higher and further, but it will also help them strengthen their home bases and reinforce their uniqueness, hence their competitiveness. Wings and roots are the two things one needs to prosper. It is true for a child; it is also true for an aerospace company.

> **KEY CONCEPT**
>
> **THE CLUSTER EFFECT**
>
> Clusters are geographic concentrations of interconnected companies and institutions in a specific field or sector.
>
> According to Michael Porter's 'Diamond Framework', four attributes of a location's microeconomic environment affect the competitiveness of its constituents:
>
> - The presence of high-quality and specialised inputs, be it human resources (e.g., qualified workers), technical resources (e.g., research laboratories, universities…) or financial resources (e.g., venture capital)
> - A context that encourages investment together with intense local rivalry
> - Pressure and insights gleaned from sophisticated local demand (e.g., Airbus in Toulouse, Boeing in Seattle…)
> - The local presence of related and supporting industries (e.g., software, electronics, information technology providers…)
>
> Clusters confer a competitive advantage that is not related to size or scale but to the location of the firm and to the linkages that exist within the cluster—often in the shape of an informal and intangible network of relationships enabled by geographical and cultural proximity.

Closing the Deal
(The Art & Craft of Selling Aerospace Products)

September 2013

I BEGAN MY CAREER in the aerospace and defense (A&D) industry as an export sales manager for a European missile manufacturer (now MBDA). It was a somewhat unusual career move for a business school graduate with no engineering background. Today, even fewer graduates have A&D high on their list when they look for jobs as sales or marketing managers. This is, after all, an industry built by engineers, and its reputation is based primarily on products and the technological innovations that underpin them. But, as we say in French, there is a long way from the cup to the lip, and a great technology does not always guarantee a great business.

The Concorde may be the most famous example of an excellent aerospace product that did not translate into commercial success, but many others can be found in an industry where engineers tend to go for the "best" solution when "good enough" would do for many customers. And when a product does thrive commercially, its success is usually attributed to innovation and technical expertise. Often overlooked are the critical roles played by a few talented and persistent sales executives.

Indeed, one peculiarity of the A&D business is that salespeople often have to evade efforts by management teams and their consultants to standardize every process and systemize every task, in order to keep doing their job properly. Sales are the part of the business where craftsmanship, personality— some would say quirkiness—and the ability to connect with a customer at a personal level can still make a difference. To the despair of many management consultants, such skills cannot be captured on a PowerPoint slide.

Seddik Belyamani, a legendary Boeing salesman, used to say that at the end of the day, "there is a real person that signs the contract, not a computer, not an Excel spreadsheet." For some, sales does not even look like a job: playing golf, eating in nice restaurants, traveling all over the world. But for those who do understand how aerospace deals are closed, it may be one of the hardest jobs on Earth.

> *Sales are the part of the business where craftsmanship, personality—some would say quirkiness—and the ability to connect with a customer at a personal level can still make a difference.*

John Leahy, Airbus's COO-customers, is said to be the most successful salesman in the history of aviation, having sold several hundred billion dollars' worth of passenger jets since he joined the company in 1985. He is often credited as the driving force behind raising Airbus market share from 18% in 1995 to more than 50% a few years later. Leahy certainly did not do it alone: Airbus's sales and marketing teams are numerous, competent, well-organized, and capable of executing deals without him. But there is no question that such individuals can have an outsized impact on the success of a business like Airbus or any other aerospace company.

On the defense side of the industry as well, competence in sales explains why some European military contractors have managed to shrug off the gloom and doom surrounding the industry. One of my clients is a world leader in night-vision technology. The fact that its products are state-of-the-art surely helps, but the company also has people in the organization, starting with the CEO, who know how to sell even when times get tough.

Sweden-based Saab has been struggling strategically for years. It is too small compared to other military contractors, too remote from the big defense markets and spread too thinly across many different businesses. Yet, year in and year out, Saab manages to survive by selling a few Gripen fighters or Combat Management Systems, and by being commercially savvy in key markets such as India or South Africa. The same can be said for top exporters like Diehl in Germany or DCNS in France.

Looking back, the years I spent selling missiles were some of the most formative of my professional life. They taught me how the A&D business really worked, lessons that I still use in my current consulting job as a reality check on my strategic recommendations. The bottom line is that a great product is worth nothing if you do not have the right people to sell it. My hope is that more young graduates from American and European business schools will choose to join some of these companies that sell cool, high-tech A&D products all over the world. These companies will need excellent – possibly quirky – salespeople over the next decades if they want to stay ahead of their competitors.

KEY CONCEPT

Big Data

Like most industries today, the aerospace community is obsessed with data, "big data" to be precise. What is it about? Simply the fact that in an increasingly digital world, data is perceived as the new source of competitive advantage. By harnessing data, billions of them, it is thought that one can not only create value – in the shape of lower operating costs for airlines, better travel experience for passengers or streamlined supply chain for manufacturers – but also capture a substantial chunk of it, provided one owns or has a privileged access to such data.

It is through that lens that recent M&A transactions such as UTC acquiring Rockwell Collins should be analyzed. It also what underpins a new type of partnerships between aerospace players and data analytics firms such as GE Aviation-Teradata and Airbus-Palantir, as well as recent investments such as UTC's Digital Accelerator or Boeing's AnalytX. All these players are in a race to capture, channel and exploit as much data about their products as possible. Data is seen as the new gold, and everybody is making a rush for it. This is the new "digital scale" game. To play and win that game, aerospace companies feel that they need to make bold and fast moves because time is of the essence, sometimes to the point of turning their organizations upside down (see column about *"Airbus' innovation gamble"*).

But this is where the industry needs to be careful. While mastering and monetizing data is clearly a huge business challenge and opportunity, it should nevertheless not let us believe that the future of aerospace belongs or should be left to data scientists only. Sure, they have a key role to play, but to grow and succeed sustainably in the digital age, companies will first and foremost need leaders who, when appropriate, are able to free themselves from the tyranny of data, and instead rely on craftsmanship, intuition and some level of practical intelligence to make the right strategic decisions.

Artificial intelligence is no substitute for practical intelligence. Knowledge (from data) is no substitute for wisdom (from experience).

Managing the Political Risk

January 2015

AVIATION WEEK'S CHOICE OF Vladimir Putin as Person of the Year for 2015 has been controversial to say the least. Among the negative comments posted on AviationWeek.com, some readers said an aviation magazine should not mingle with politics. Yet one of the key features of the aerospace and defense (A&D) industry is its close link to international politics. Trying to analyze and report on the industry while ignoring this dimension would be like missing the forest for the trees.

Selling airplanes, warships, satellites, or weapon systems has never been only about price, product performance or technology, and it certainly does not happen in a political vacuum. The sale of helicopter carriers by France to Russia in 2011 in a $1 billion deal was controversial from the start. Complaints came from the US as well as from NATO. The deal was the first sale of military technology to Russia from a NATO member state. Yet the most surprising element was not that France sold to Russia—all defense-exporting countries are desperate for orders—but rather that Russia bought from a NATO country. It says something about the state of the Russian shipbuilding industry.

When it comes to major international defense contracts, politics often trump economics.

It is therefore no shock that when the relationship between Russia and Europe went sour over Ukraine, the French government decided to suspend the delivery of the first warship, which was due in November last year. This is a major snub for Putin, and it is already rumored that Russia is now trying hard to derail France's longstanding effort to sell the Dassault Rafale to India in retaliation and offer the Sukhoi Su-30 fighter jet instead.

But despite appearances, this is business as usual for A&D players. When it comes to major international defense contracts, politics often trump economics, which is why, for example, countries like South Korea and Japan have always bought American when it came to strategic assets such as fighter jets and antimissile defense, often in spite of the economics of the deal.

This is also why European defense companies try hard to develop and market their products as "ITAR-free," to avoid being subject to US

International Traffic in Arms Regulations (ITAR[ii]), which govern exports of militarily sensitive technologies and give the US administration a say in every sale of military systems that contain American components. As a case at hand, a French contract to sell spy satellites to the United Arab Emirates was stalled for more than a year because the spacecraft carried US components subject to ITAR. It took several months of negotiation and a deal brokered by Presidents Barack Obama and Francois Hollande to resolve the issue. The takeaway for companies involved in exporting sensitive products is they should consider politics as not just an externality over which they have no control but as important input into their contract negotiations as well as their business strategy as a whole.

I remember negotiating a major missile contract with an Asian government, for which the "force majeure" clause became a major stumbling block as the client did not want to be at the mercy of a political embargo, even in disguise. When you consider the criteria for an event to be qualified as "force majeure" (under French law, there are three main criteria: externality, unpredictability, and irresistibility), it was quite tricky to agree on what would be a "force majeure" event in that context.

Beyond specific contract negotiations, it is also important for companies to incorporate the political factor into their overall business strategy. Scenario-building and "wargaming" are two efficient strategic tools to use. Although it is just simulation, it can lead to powerful insights into the likelihood of various political events and can thus inform a business strategy by laying out possible sequences of events leading to different outcomes.

[ii] **International Traffic in Arms Regulations (ITAR)** is part of the US export control regime. It regulates US origin items (including items initially made abroad but coming into the US) specifically designed, modified, configured or adapted for military and space applications, as well as, related technical data and services. In general, ITAR controls the export, re-export, re-transfer and temporary import of defence articles (including technical data) and the provision of technical assistance. An export includes not only physical exports, but also oral and visual disclosures of technical data to a foreign person.

The ITAR is therefore one of the broadest and most stringent export control regulations in the world, due to the fact that the regulations control not only major military equipment but similarly control innocuous components, parts, accessories and attachments if specifically designed, developed, configured or adapted for a military or space application. Under its 'See Through' rule, an article that is subject to the ITAR will retain its ITAR status even when incorporated into a higher-level platform. This remains true even when the higher-level article is designed for civilian or commercial end-use.

ITAR implies several constraints for customers around the world. In particular, if and once the authorization for exportation is granted, the product remains subject to ITAR, which means that US Government's approval is required for any re-export or transfer to other countries. As an illustration, in 2018, the sale of the Rafale aircraft by France to Egypt was blocked by the US Government because one of the components (an electronic chip) found in the SCALP missile (sold with the Rafale) was US-originated. That is why many European companies advertise their products as "ITAR-free" (i.e., with no US parts inside), which means that they will not be subject to US extraterritorial oversight.

Ultimately though, the best way to develop a strategy resilient to international politics is to create business interdependencies that will make damaging political decisions by one country more difficult to implement. Be it through an industrial joint-venture, a technology licensing agreement, or other forms of partnership, the more balanced the stakes, the less likely a deal-breaking political decision will occur.

In this environment of political instability, scarce resources and shrinking defense budgets, the best way to make one's business more resilient to political uncertainties is to move away from pure win-lose competitive situations. A modern winning strategy is not only about sharing the economic pie, but about protecting it from political interference and—whenever possible—increasing its size. "Co-opetition" (i.e., preserving or increasing the pie while sharing it), not competition, is the new name of the game.

> *The best way to make one's business more resilient to political uncertainties is to move away from pure win-lose competitive situations.*

KEY CONCEPT

Co-opetition

Using the metaphor of a pie to describe a market, one can say that any business activity has two different aspects: the creation of a pie and the division of a pie. The former is achieved through cooperation, the latter through competition.

The concept of "co-opetition", pioneered by business school professors Adam Brandenburger and Barry Nalebuf, is meant to integrate these two dimensions of cooperation and competition in business strategy. It implies that in a given market, relationships between players are never purely competitive or cooperative, but a mix of the two. The players are "complementors" in creating the pie (or increasing its size) and they are competitors in dividing the pie.

Let us take the example of an Asian Government intending to buy military aerospace equipment from a Western country, but only if it comes with some significant industrial offsets and technology transfer.

In such context, a local company which would traditionally be seen as a competitor by the Western supplier, may become a complementor if, by cooperating with them through some kind of offset or technology transfer agreement, it allows the Western company to break through the local market. Similarly, by cooperating with the Western supplier, the local company will grow its business in a way that would not have been possible otherwise.

Such co-opetition dynamics create inter-dependencies between players that, once established, will make it more difficult for a government to break a deal for political reasons.

Chapter 5

GROWTH STRATEGIES

"For every activity, there is an appropriate scale. Today we suffer from an almost universal idolatry of giantism. It is therefore necessary to insist on the virtues of smallness."

E.F. Schumacher, German economist (1911–1977)

CHAPTER INTRODUCTION

This chapter captures two key ingredients of corporate strategy: dynamic business portfolio management (selling and buying business units in a way that significantly changes the overall business perimeter) and Mergers & Acquisitions (M&A).

Dynamic business portfolio management is particularly relevant to corporate growth strategy because it is hard to trigger a new growth phase without changing the business mix in the first place. This means not only investing in new businesses but also divesting activities that are sub-scale, not exploited to their full potential or constrained by being part of the group.

This topic was already touched upon in the "Industry Structure and Competition" chapter, from the point of view of both the US and the European industry. Here we look at it through the lens of three specific companies: the first one – Thales – which does not practice it when it should, the second one – General Dynamics – which practiced it in the most dramatic way, and the third one – Airbus – which, after resisting it for years, finally made the jump by divesting a significant part of its defense portfolio, possibly showing the way for other European A&D groups.

Mergers & Acquisitions (M&A) is part of the standard corporate strategy toolkit of modern corporations. The chapter covers two high-profile merger cases (among many others) that happened in the A&D industry: The first one is the 2015 merger of Harris Corporation with Exelis (formerly ITT). The second one is the 2018 Safran-Zodiac merger. In both cases, the strategic value creation logic for the deal was far from obvious but, as often in this type of transactions, many other factors came into play to make them happen.

Shrinking to Grow

November 2013

SINCE HE WAS APPOINTED CEO of Thales earlier this year, Jean-Bernard Levy has repeatedly proclaimed that there is nothing wrong with the company's business portfolio, because all of its operational units are profitable, if only marginally. Unfortunately, this view is symptomatic of the way most European aerospace and defense (A&D) players fail to understand the value of dynamic business portfolio management. Indeed, the traditional view—derived from deep-rooted vertical integration and a controlling mindset—is that businesses should only be sold when they lose money, or when cash is urgently needed. In the case of Thales, though, there seems to be a strong rationale to rethink its portfolio, regardless.

The group's portfolio is indeed complex, confusing and inconsistent, the result of chaotic historical growth and typical French-style industrial engineering. Starting with the takeover of Sextant Avionique and the grabbing of various assets from Dassault Electronique, Aerospatiale, and Alcatel in the 1990s and extending to the acquisition of a 35% stake in DCNS, Thales has become a "five-legged sheep," present on all steps of the value chain—from components to systems, from mortars to satellites—but with stagnant sales and a stock price that has been lagging for several years (*see graph below*).

It is hard to imagine how the company's management can expect to trigger a new wave of growth without changing its business mix. This means not only investing in new businesses but also divesting activities that are sub-scale, not exploited to their full potential or constrained by being part of the group. Unfortunately, there are plenty of such languishing businesses in large European A&D companies.

In the US, dynamic portfolio management has been part of the A&D players' strategic toolkit for at least two decades.

That contrasts with the US, where dynamic portfolio management has been part of the A&D players' strategic toolkit for at least two decades. The pioneer was General Dynamics (GD), which radically transformed itself in the 1990s to become a stellar performer. GD's transformation started in the early '90s with the disposal of three of its main businesses: The F-16 to Lockheed Martin, tactical missiles to Hughes and Cessna to Textron. At that point GD was essentially left with two core businesses: its nuclear submarines and the M1 battle tank, with 24,000 employees compared to 80,000 three years earlier.

Next, GD went on an investment spree, spending $15 billion to acquire around 30 companies between 1995 and 2003, which led to a completely new group, well-positioned in such diverse businesses as C4ISR, business aviation, information technology services and combat systems. In fact, the ability to shape and manage the portfolio in a dynamic way—based on consistent and transparent criteria such as return on net assets and cash return on invested capital (ROIC)—became one of the core capabilities under the long-term leadership of Nicholas Chabraja (in that respect, stability and continuity does help).

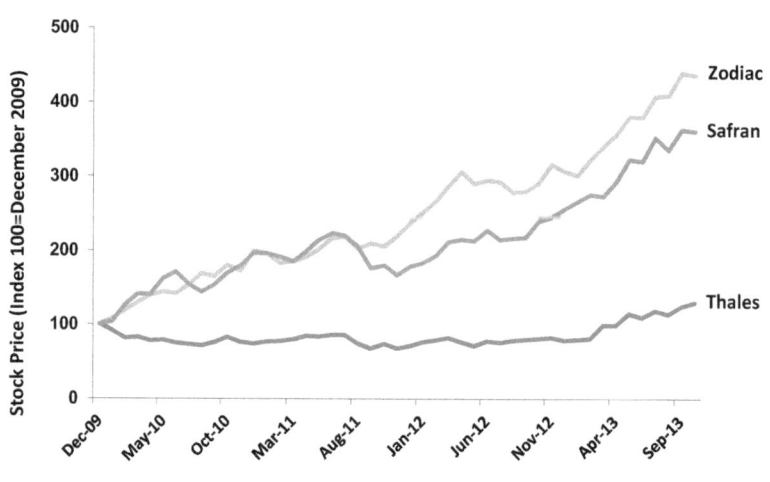

THALES STOCK PERFORMANCE VS. FRENCH PEERS 2010–2013

Source: Yahoo! Finance

Other leading US players have applied similar, if not as drastic divestment strategies to shape their portfolios: Raytheon spun off about $6 billion worth of businesses (almost 20% of its portfolio) in the early 2000s. Lockheed Martin, Boeing and Northrop Grumman also collectively disposed of close to $10 billion worth of businesses in the same period.

Of course, portfolio management is not the only ingredient of successful performance. But the capacity to change course and catch better winds is a core element of any well-advised navigation strategy, especially when the existing winds are not so strong, as is the case in the defense sector.

Similarly, the ability to let go of some historical baggage is often a healthy first step toward a new growth path. For European A&D companies, such baggage is typically quite heavy as most groups have been built through successive waves of conglomeration, often based on political grounds.

> *The ability to let go of some historical baggage is often a healthy first step toward a new growth path.*

So could European companies emulate their US counterparts by divesting major chunks of their portfolios on their paths to new growth phases? EADS, Safran and Thales have all recently been offloading some small, non-core businesses. But the kind of divestment or spin-off that would really "move the lines" of their portfolios and of the industry's competitive dynamics needs to be much more significant, possibly involving some activities still considered "sacred cows." That may be more difficult than effecting a merger or acquisition. For some CEOs, "divesting" seems to be the hardest word.

KEY CONCEPT

BUSINESS PORTFOLIO MANAGEMENT

Business portfolio management (i.e., acquiring and/or divesting individual businesses) is one of the core functions of a corporation's leadership. It relates to the choice of a scope, which is itself one of the most important corporate strategic decisions. It also relates to whether owning a specific business creates a corporate advantage. This translates into three fundamental questions, that the management should address regularly:

1. Does ownership of the business create benefit somewhere in the corporation?
2. Are these benefits greater than the cost of owning it (i.e., corporate overhead costs)?
3. Does the corporation create more value than any other possible corporate parent or alternative governance structure for the business?

Over time, business portfolio management decisions (acquisitions or divestments) are driven by three interconnected factors: the firm's external environment, the overall corporate portfolio performance and trajectory, and each individual business' performance and trajectory. Since these factors are constantly changing, it is important to keep monitoring them and manage the business portfolio accordingly.

Airbus Shows the Way

October 2014

THE ANNOUNCEMENT BY AIRBUS GROUP a couple of weeks ago that it was putting €2 billion ($2.52 billion) in defense and space businesses up for sale, with another €1.5 billion worth of businesses (defense electronics) tagged as question marks still to be addressed, is an important milestone not only for Airbus itself but also for the European aerospace and defense industry.

> *Historically, European groups have had a very conservative approach to divestments: They would only consider selling money-losing businesses.*

It may well be the first time a divestment program has been launched by a large European group so proactively. Historically, European groups have had a very conservative approach to divestments: They would only consider selling money-losing businesses. Such divestments were seen as "last resort" options and not meant to create value for the seller.

One reason why divestments have not been more popular in Europe is they can end up being expensive and disruptive for the seller. It is well known that European labor regulations make it hard for companies to hire and fire people. But they also make it hard to buy and sell businesses.

Trade unions—in France, in particular—almost automatically call for strikes as soon as a divestment plan is announced, triggering a long and expensive process of negotiations to safeguard some benefits or compensation for the workforce. This is symptomatic of a pervasive mindset among Europeans: They always worry about what they are going to lose before even knowing what they might gain with a change of ownership. Most workers are made to believe their jobs are safer and ultimately, they will be better off if they remain in a big group.

History shows that is far from true. Many small business units within large groups can be neglected and undermanaged. Without proper investment, they lose their competitive edge. They also tend to be burdened with high overhead to pay for the bureaucracy of the mother organization. Finally, they have to follow the rules of large organizations in terms of wage minimums and social benefits, even though they might essentially be running as small companies and competing with other small firms that are not subject to such constraints, putting them at a competitive disadvantage.

Last year, as a one-off divestment, Airbus Group decided to sell a small business called Test & Services [now Spherea]. This 500-employee company is a world leader in electronic test benches for the aerospace industry. As soon as the sale was announced, trade unions raised fears that such a transaction would be detrimental to the company. The reality rather was that ownership by a big, vertically integrated group had been detrimental to Test & Services for many years and it was a matter of survival to spin off the business. Six or seven years ago, as a subsidiary managed at arm's length, Test & Services had annual sales of more than €130 million, its profit margins consistently topped 10%, and its business plan put it on a course to reach €400 million in sales within five years.

> *Many small business units within large groups can be neglected and undermanaged. Without proper investment, they lose their competitive edge.*

But after an initial reorganization that led to its full integration with other defense electronics businesses within the Airbus Group, Test & Services started losing its competitive edge, and its revenues and profitability declined sharply. In 2012, its revenues were down to less than €90 million and its valuation was a fraction of what it had been six years earlier.

The bottom line is that not all businesses are equal in a large group. Whether a business benefits or suffers from such ownership depends on multiple factors, such as: the competitive dynamics of its market, potential synergies with other parts of the group, position of the business in its life cycle (startup versus mature) and its alignment with the group's overall strategy. And things can change over time, hence the need for dynamic business portfolio management and a systematic process for regularly reviewing and assessing the benefits of owning a specific business at a certain time.

In that respect, Airbus Group's divestment plan is to be praised. It sends a strong signal that no business within the group is safe unless it is identified as being core and it has a clear growth path. The plan also reinforces Chief Executive Tom Enders' message that the culture of the group needs to change, with a stronger focus on financial value creation—including divestments of businesses that may be profitable but may not be best positioned to create further value within the group.

Let us hope other European groups will follow, as this might finally trigger the major reconfiguration that the region's A&D industry has been longing for.

[UPDATE: Airbus' defense electronics business (€1 billion of revenues, 4,000 people) was sold to US private equity firm KKR in 2016 for €1.1 billion and was renamed Hensoldt. It was less than half of the initial divestment plan, but it remains the largest spin-off by a European group to date.]

MERGER LOGIC
(HARRIS-EXELIS)

MARCH 2015

MERGERS AND ACQUISITIONS HAVE an interesting side effect: They often bring to the surface things that otherwise would remain hidden or unnoticed. The recently announced merger of Harris Corp. and Exelis is a case in point. Beyond the usual M&A lingo ("transformational," "powerful combination," "growth platform," "scale" etc.), this transaction brings back to light some fundamental truths about the US defense sector that otherwise tend to be forgotten.

The first truth is that it has become increasingly hard to sort out what—in this type of transactions—is financial engineering from what is industrial engineering. After all, we are in the era of financial capitalism, and industry often has become a mere background or raw material to financially motivated deals. This is not to say this is the case for the Harris-Exelis transaction, but the business story does not look as compelling as the financial one.

> *It has become increasingly hard to sort out what is financial engineering from what is industrial engineering.*

The reality is that Harris and Exelis have experienced declining revenues for several years, and that merging them can be seen as an artificial way to boost their bottom lines, without real growth behind it. Remember that Exelis was spun off from ITT only four years ago and that part of it was made into Vectrus Inc. last year. On the face of it, merging Harris and Exelis is a growth story since Harris's revenues will soar to $8 billion from $5 billion today. However, a back-of-the-envelope analysis shows that the aggregate revenues for these companies have been declining slowly but surely for the last five years, to $9.8 billion in 2014 from $11.5 billion in 2009 including Vectrus (*see graph below*).

In that respect, this transaction may be seen as much as a reflection of the state of the US defense industry—struggling to cope with post-war budgets—as an attempt to rebuild some superficial strength in the defense electronics and communications sector, like taking several pieces of leftover bar soap to make a new one.

The second truth is that the US defense market is both a fortress and a jail. It is a fortress because, despite years of talks about globalization and transatlantic consolidation, the US defense market remains inaccessible to foreign companies, and all significant M&A transactions have been among US players only, the Harris-Exelis merger being the latest one.

The story of BAE Systems is often quoted as a success for transatlantic consolidation, but the reality is that it forced BAE to run itself as two separate companies with little synergy between the two sides, with damage to the overall company's management and strategic direction. The other major attempt by a foreigner to break through the US defense market was Finmeccanica's acquisition of DRS Technologies and that was a fiasco.

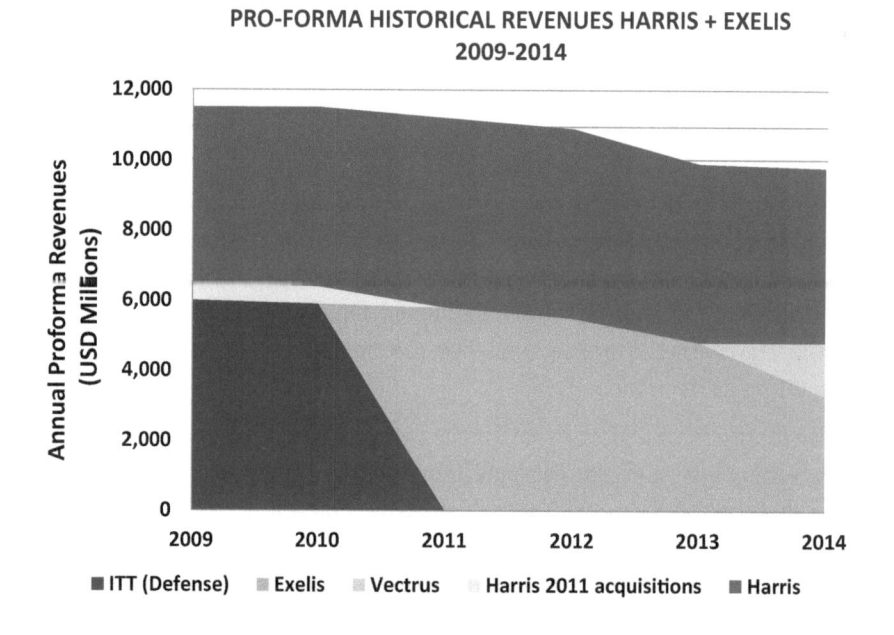

PRO-FORMA HISTORICAL REVENUES HARRIS + EXELIS 2009-2014

Source: Company Annual Reports

The flip side of this "Fortress USA." for domestic players is that their market, while inaccessible to outsiders, is also a jail for themselves from which it is hard to escape. Indeed, when a business is so outrageously dominated by one main customer—in this instance the US government—it creates an invisible barrier to diversifying the customer base internationally.

The competitive dynamics and routes to market are so different that it takes a new set of skills, organization and even culture to succeed in a meaningful way, and it is almost impossible to build on the back of such a dominant and specific customer. Despite repeating year after year that internationalization was part of their strategy, Harris Corp.'s and Exelis' international revenues (excluding foreign military sales) were only 9% and 11% of their total revenues last year,

> *Sometimes, when companies struggle to revive their top-line growth in a tough market environment, the only way forward is to reshuffle the cards and perform a "strategic reframing."*

respectively. Overall, merging these two companies just reinforces the message that the US defense market is "for American eyes only" and that—like in the World Series—US players do not need to compete internationally to call themselves world champions.

The final truth is that, sometimes, when companies struggle to revive their top-line growth in a tough market environment, the only way forward is to reshuffle the cards and perform a "strategic reframing." This can be accomplished through either a change of ownership, a reconfiguration of the business portfolio (by combining separate businesses and/or by spinning off others) or a redefinition of the company's playing field. One could interpret the Harris-Exelis merger as such an attempt to reshuffle the cards, creating a new home for Exelis and reconfiguring the new business portfolio in a more synergetic and innovative way.

This is by far the most positive spin we can put on it. Otherwise, we may have to look at the merger with a bit of cynicism and worry that it is another example of a nicely packaged, financially engineered zero-sum deal.

Hope and Despair
(Safran-Zodiac)

April 2017

IN JANUARY, AEROSPACE SUPPLIER SAFRAN announced it had launched a $10 billion friendly takeover of fellow supplier Zodiac Aerospace, a world leader in aircraft cabin and safety equipment.

In the wake of the announcement, TCI, a UK-based hedge fund and minority shareholder in both Safran and Zodiac, publicly expressed concerns about the merger, criticizing its strategic and financial rationale as well as the way it is being forced upon public shareholders, calling it "unfair, unscrupulous and unbecoming" of such a company.

> *Mergers and acquisitions are an integral part of modern capitalism. It reminds us that a company cannot rest on its laurels and that if management is too complacent, it gets punished.*

As Safran's board and TCI exchange blows, are we seeing modern capitalism at its best or worst? I tend to think it is a bit of both. There is nothing wrong with the takeover attempt itself. Mergers and acquisitions are an integral part of modern capitalism. It reminds us that a company cannot rest on its laurels and that if management is too complacent, as Zodiac's was in recent years, it gets punished. By underperforming, it opened itself up to predators. And Safran had been waiting for its prey for many years.

Neither is there anything wrong with TCI making the dispute public and, by doing so, lifting the veil of secrecy that often surrounds this type of transaction. As listed companies, both Safran and Zodiac are expected to disclose enough information to allow for healthy scrutiny of the deal and more transparency toward all their stakeholders.

Capitalism also is at its best when companies use transactional mechanisms to move business boundaries, thereby stimulating innovation and developing synergies that ultimately benefit customers. In this case, that could occur in the field of electric power, where Zodiac has strong expertise and Safran has much at stake.

Unfortunately, the clash between Safran and TCI is also capitalism at its worst because so much pretense and self-interest are on display by all parties. TCI presents itself as the defender of small public shareholders, but hedge funds are notorious for creating value for themselves through ruthless tactics. Meanwhile, Zodiac's shareholders six years ago rejected a similar offer from Safran, arguing that a merger would bring no synergies and the two companies' cultures were incompatible. But a two-year rough patch was enough to make Zodiac's board and controlling families cave in, with complete disregard for the values that made the company so successful for decades: entrepreneurial spirit, humility and independence.

Finally, Safran is trying to make everyone believe this transaction will create a lot of value when all evidence points to the contrary. The only area of real positive synergy—electric power generation—accounts for less than 10% of Zodiac's revenues and hardly justifies acquiring the whole company. Also, their ability to increase their "share of wallet" with their main customers by combining the two companies' product portfolios is questionable, as both companies are already individually more dominant in their respective domains than these customers probably would like.

Over its 14 years in existence, Safran has always struggled to create value beyond its core legacy Snecma business (mainly aircraft engines), which has consistently accounted for more than 90% of the group's total profits (*see graph*). The 2003 merger of Snecma with Sagem that created Safran was based on similar grounds to the one with Zodiac, yet the announced benefits never materialized, and most Sagem businesses were sold off within a few years. Similarly, in 2009, Safran invested in the security sector, successively buying GE Security and L-1, brandishing it as the "third pillar" of the group, with an ambitious 10-15-year growth target. Six years later, it was put up for sale. For an industrial operator supposedly keen on long-term value creation, this stint in security comes across as rather short-sighted and ill-advised.

SHARE OF LEGACY SNECMA BUSINESS IN SAFRAN PROFITS

- Legacy Snecma Business - Operating Profit
- Other (Acquired) Business Lines - Operating Profit
- Share of Legacy Snecma Business in Safran Profits

Year	2003	2004	2005	2006	2007	2008	2009	2010	2011	2012	2013	2014	2015	2016
Share of legacy Snecma business in total Safran Revenues	67%	65%	66%	68%	75%	85%	81%	79%	78%	79%	81%	82%	82%	92%

Disposal of Sagem businesses

Security acquisitions

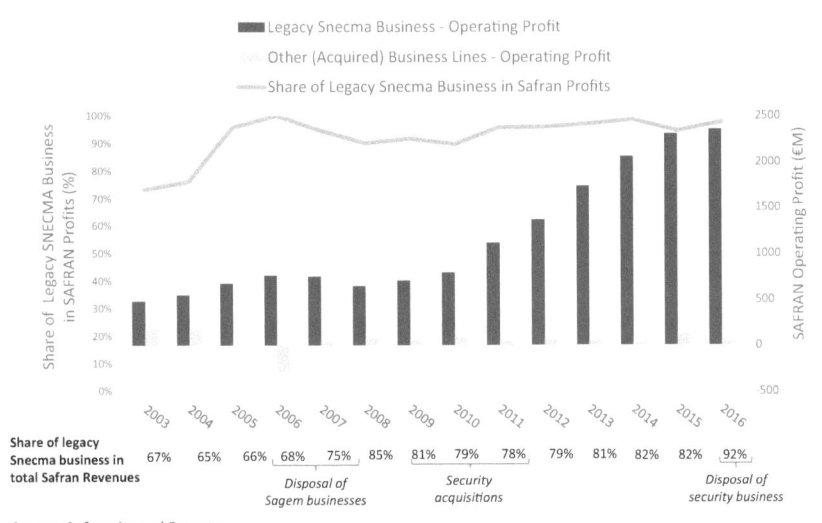

Disposal of security business

Source: Safran Annual Reports

Will Zodiac suffer a similar fate? All in all, if the deal goes through despite its flaws, it should be greeted with a mix of hope and despair. Hope to see people on both sides use their creativity and goodwill to meet the challenge of making the combined company worth more than the sum of its parts. Despair over noting that the fate of companies, in aerospace as elsewhere, is too often driven by egos and short-term self-interest rather than by true stewardship and strategic vision.

Chapter 6

CORPORATE STORIES

"We don't clone divisions... If you're in California and you wear sandals, you wear sandals."

Frank Lanza, co-founder L-3 Communications (1931–2006)

CHAPTER INTRODUCTION

This chapter is a collection of case studies on the successes and failures of iconic A&D companies: BAE Systems, Dassault Aviation, L-3 Communications, Airbus, Bombardier, and Rolls-Royce. In each case, key strategic challenges and decisions are highlighted, analyzed, and commented.

In the case of **BAE Systems**, the overall feeling is one of a company that has run out of steam and is critically short of strategic options, following several questionable decisions made over the years.

Dassault Aviation comes across as an anomaly in the industry but there is something uplifting about an organization that so consistently manages to defy the odds, challenges commonplace analysis and so expertly balances craftsmanship and high-tech, as well as conservatism and innovation.

The story of **L-3 Communications** offers a fascinating, condensed version of the life of a corporation from cradle to grave. After being an incredible success story, thanks to a beautifully crafted and perfectly executed strategy, it slowly but surely lost its magic to the point where it had no choice but be acquired or merge with another company.

Airbus's 50th anniversary in 2019 made it a perfect time to reflect on the state of the group. While a lot has been achieved in 50 years, one cannot look at the group's results without getting a sense of waste and fragility.

The next column explains why **Bombardier**'s tenure in commercial aerospace – that ended in 2019 with the sale of its CRJ business - will ultimately be remembered as a financial and strategic failure.

The final column reflects on the recent downfall of **Rolls-Royce** and the possible reasons of its current misfortunes.

OUT OF STEAM
(BAE SYSTEMS)

JULY 2014

FARNBOROUGH IS BAE SYSTEMS (BAE)' home turf, and the air show, which takes place there next week, has traditionally provided the perfect stage for the company to show off its products, political clout, commercial successes and strategic initiatives. Indeed, over the last 15 years, BAE has been the frontrunner of the aerospace and defense (A&D) industry when it came to making bold strategic decisions: the merger with GEC-Marconi in 1999, investing massively in the US between 2000 and 2007, creating a defense electronics joint venture with Finmeccanica in 2003, exiting the commercial aerospace business in 2006 and attempting to merge with EADS in 2012.

Yet the overall feeling surrounding BAE today is one of a company that has run out of steam and is critically short of strategic options. In fact, several of its decisions seem to have backfired. They include the 2006 sale of its stake in Airbus (a business for which it was eager to pay a premium to reenter by 2012), the divestment of its share of Saab in 2010–11 (Saab's market cap has doubled since then), and its multiple acquisitions in land systems, which is now one of the most distressed sectors in the defense industry.

> *The overall feeling surrounding BAE today is one of a company that has run out of steam and is critically short of strategic options.*

This dire strategic position is even more striking when compared with other UK A&D players such as Meggitt and Babcock International, which have been riding high in similarly difficult market conditions (*see graph below*).

One might therefore wonder if all the strategic moves made by BAE were as smart as they were bold. Indeed, a very rough analysis shows that between 2000 and 2014, the company's revenues have increased by almost exactly as much as the total of its acquisitions (net of disposals)—that is around £6 billion ($10.3 billion). This tends to suggest the value created from all these acquisitions has been limited. The company has "captured" a lot of value, but has it actually "created" value?

In this regard, I will limit myself to two comments. First, value creation usually occurs when one fulfills a customer's untapped need. This process typically requires a company to significantly transform its underlying capabilities and the way it interacts with its clients. For example, the story of VT Group between 1996 and 2008 (before its breakup), moving away from its historical shipbuilding business and becoming a leader in outsourced government services, illustrates a true value-creation strategy.

BAE SYSTEMS STOCK PERFORMANCE VS. MEGGITT AND BABCOCK INT'L JULY 2009 - JUNE 2014

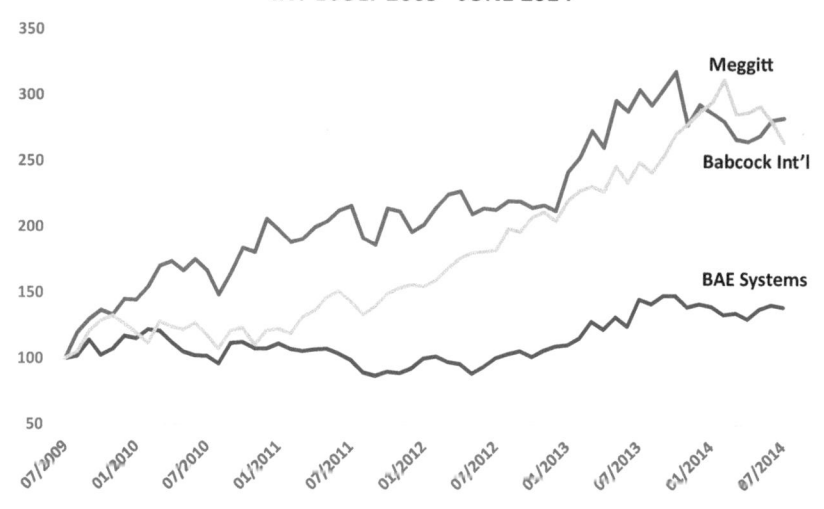

Source: Yahoo! Finance

As far as BAE is concerned, there has been no such story. While there were high hopes that the British Defense Industrial Strategy of the mid-2000s would trigger a breakthrough value creation process for both the Ministry of Defense and BAE, its implementation did not live up to its promise.

Second, creating value is not just a matter of becoming bigger. Of course, economies of scale are important in a highly capital-intensive industry but, by definition, economies of scale happen when a single product is manufactured in great quantities. It does not work when multiple products are made in relatively small quantities. That is, by and large, what BAE has been doing. Its top-line growth does not seem to have translated into value creation, as the economics of the business have remained essentially unchanged.

So what are the options for BAE? Like any other business, it must choose between two strategic postures: ride an existing wave or create its own. So far, BAE's strategy has consisted of riding successive waves. It rode the Airbus wave until the winds started to shift, then it moved on to the "US defense bonanza" and became a "pure play" US-centric defense player, then it jumped onto the security wave, rebranding itself as a "defense and security" company for a while. The proposed 2012 merger with EADS was an attempt at remounting the commercial aerospace wave. Today, there is no obvious wave to catch and the company is at loose ends. Therefore, the only way forward is for BAE to create its own waves.

> *It is possible that in its current configuration BAE has reached the end of its journey, and that the only way to create value is to reshuffle its business portfolio, possibly breaking up the company in the process.*

Indeed, it is possible that in its current configuration BAE has reached the end of its journey, and that the only way to create value is to reshuffle its business portfolio, possibly breaking up the company in the process. In doing so, it could rediscover the frontrunner status that made it famous in the first place and plant the seeds for a new "star of the show" to emerge at Farnborough. Until then, we will have to make do with a lackluster performer.

Dassault's Secret Sauce

May 2015

IN THIS COLUMN LAST YEAR, I wrote "[Dassault Aviation's] Rafale is a story of stubbornness and manipulation," noting that 28 years after its official launch, the aircraft had not won a single export order. But within the last two months, three orders have been announced for close to a total of 100 aircraft to Egypt, India, and Qatar, and suddenly the future looks bright. So what's happening?

Yes, Rafale is a story of stubbornness, but sometimes it pays off. Rafale also is a story of manipulation, but one must acknowledge that political manipulation is part of the game.

And so these recent commercial successes remind us of the singularity of the international defense market: unpredictable, highly political, and with economic and industrial stakes so big that it can take a decade or more to close a deal.

Everything that has been happening to the company during the past half-century seems to defy rational analysis and common business sense.

It also reminds us of the singularity of Dassault as a corporation. Everything that has been happening to the company during the past half-century seems to defy rational analysis and common business sense: It took 30 years for the company to sign its first export order for the Rafale, even though one of the main design drivers of the aircraft was to make it light enough to be exportable (unlike the much heavier Eurofighter).

Which CEO today would dare to show up in front of his shareholders and present a business plan in which the first export order would only happen a couple of decades down the line?

Meanwhile, compared to its giant American counterparts such as Boeing or Lockheed Martin, Dassault—with its meager $5 billion of annual revenues— looks incredibly small for an aircraft OEM capable of designing and producing what is widely recognized as the best alternative to US combat aircraft. And in spite of all the industry consolidation in Europe during the last two decades, Dassault has remained fiercely independent and barely grown in size. Yet the company is consistently profitable, thanks primarily to a tight financial discipline and its Falcon executive jet business, which accounts for almost three-quarters of Dassault's annual sales today.

There is certainly something anachronous about the company: the level of secrecy it cultivates, the way it seems to go against the tide of globalization and consolidation, and the way it is able to rise from the ashes as the recent export orders tend to indicate.

The mystery and fascination surrounding Dassault is not new. In a 1973 Rand Corporation report[iii] commissioned by the US Air Force, Robert Perry writes extensively about the company's paradox. He wrote: "That Dassault is consistently able to create and produce high- performance aircraft comparable to and competitive with those of the United States is almost paradoxical, given the resources of the company and the international environment in which it operates."

Trying to come up with some rational explanation for such success, he concluded: "In many respects, the uniqueness of Dassault appears to be explainable mostly in terms of the company's people, principles, policies and practices."

What is certain is that behind what may seem old-fashioned or unorthodox practices, the company has always been at the forefront of technology, thanks to a consistent and evolutionary approach to innovation. This has allowed Dassault to capitalize on every prototype built; and it has built a lot of them.

These days, people make a big deal out of data analytics but it seems Dassault has been harnessing the power of data for many decades, as its heavy reliance on prototype building has given the company unprecedented insights into what it takes to develop and build a fighter aircraft. For that reason, Dassault's cost estimates are believed to be very precise, with no more than a 10% margin of error, compared with US programs' typical 40–80%.

Overall, it would be easy to characterize Dassault's recent commercial revival as a stroke of luck and possibly the last stand of a company that has been living off a decades-old business philosophy and a network of political and commercial connections whose relevance and power have been weakening slowly. Blame the forces of globalization and modern capitalism.

There is something uplifting about an organization that so consistently manages to defy the odds, challenge commonplace analysis and so expertly balance craftsmanship and high-tech, conservatism and innovation.

[iii] **Rand Corporation: A Dassault Dossier:** Aircraft Acquisition in France, by Robert L. Perry (1973) https://www.rand.org/pubs/reports/R1148.html

Yet there is something uplifting about an organization that so consistently manages to defy the odds, challenge commonplace analysis and so expertly balance craftsmanship and high-tech, conservatism and innovation. Dassault might even give us a clue about how to develop a sustainable corporate model in the 21st century.

After all there is a lot to learn from outliers, and possibly a lot to gain from being one. If we only try to replicate what the majority does, we will remain merely average. And average is the one thing a combat aircraft cannot afford to be.

How L3 Lost its Magic

December 2018

THE RECENTLY ANNOUNCED MERGER of Harris Corporation and L3 Technologies marks the end of an extraordinary story that has largely shaped the US defense industry over the last two decades. It also offers a condensed version of the life of a corporation from cradle to grave.

For corporate strategists, L-3 Communications, as it was called at the time of its inception in 1998, was a dream come true: a beautifully crafted and perfectly executed strategy making sense of a changing demand and supply environment to create a unique value proposition.

> *For corporate strategists, L-3 Communications was a dream come true: a beautifully crafted and perfectly executed strategy making sense of a changing demand and supply environment to create a unique value proposition.*

The co-founders—CEO Frank Lanza and Chief Financial Officer Robert La Penta—saw what was happening at the top and bottom of the defense industry pyramid and created a market space for themselves in between, at "mezzanine" level as they coined it. They literally turned the industry upside down by deploying a horizontal integration strategy (putting together a portfolio of tens of loosely connected businesses), thus generating huge economies of scope, at a time when vertical integration and economies of scale were still the predominant industry paradigm. It was "New Defense" decades before "New Space."

L3 became eponymous with a new independent, merchant supplier model which consisted, in Lanza's words, in "building and developing 'boxes' [or functional modules] as opposed to airplanes, tanks and ships" and offering them to multiple bidders on prime contract bids. It was such a smart and unique model that, in Europe, many insiders, including myself, were advocating the creation of a "European L3."

If the strategy was brilliant, the execution itself was flawless, at least for a while: Within the first 10 years, L-3 acquired close to 100 companies, all with similar characteristics ("technologically sophisticated companies with good products but that lack the resources or management expertise to grow") and all were integrated in the same way, as explained by Lanza. "We let the companies we buy keep their autonomy," he says. "We centralize marketing, coordinate research and development and give them political help. But we do not micromanage, and we don't clone divisions . . . If you're in California and you wear sandals, you wear sandals."

Thus by 2008, L3 had grown from $800 million to $15 billion of revenues and from 8,000 to 65,000 employees. L3's portfolio became the "Sears catalog" of the US defense industry, with thousands of products on offer. But the model that worked so well for a while reached its limits. Once such a company reaches a certain size, the cost of complexity overrides whatever economies of scope it may achieve in theory. Instead of critical mass, it ends up with a critical mess, and the only way to make it work is to go after fewer but larger contracts, which means moving up the value chain, which in turn means losing what made it unique and valuable compared to the "gorillas" (as Lanza used to call the big players at the top of the pyramid). This is exactly what L3 did, and it was the beginning of the end.

This is not to say that the company was badly managed after 2008, but from then on, it lost its entrepreneurial spirit, agility and—some would say—cockiness, and that was reflected in its results ever after (*see graphs below*). Even as it downsized to $9 billion in revenues in 2015 from $15 billion, it never got its mojo back and by then, there were only two ways forward: either break up the company or be acquired by or merge with another company to create another "gorilla," which is what just happened.

A well-crafted and well-executed strategy does pay off and being able to anticipate changes in the environment is essential to carving out a unique niche in which to capture a disproportionate share of the market.

So what lessons can be learned from this exceptional but relatively short-lived story? First, a well-crafted and well-executed strategy does pay off, and being able to anticipate changes in the environment is essential to carving out a unique niche in which to capture a disproportionate share of the market. Second, vertical integration and economies of scale are not the only way to create value in the defense industry.

Above all, L3's story is a reminder that every organization goes through cycles similar to the four seasons: spring (burgeoning, fast-growing), summer (blossoming, showing full strength), fall (slowing down, weakening), and winter (contracting, freezing). The key is to acknowledge the start of the fall season and begin regenerating before winter comes, in preparation for the next cycle.

Many companies fail to do that. Instead, they keep basking in their summer glory, and by the time winter comes, it is too late to change, and they

CORPORATE STORIES

are stuck, for a very long time . . . until, one way or another, they are no more. R.I.P. L3.

1998-2008: L-3 STOCK PRICE VS. DJI INDEX AND PEERS

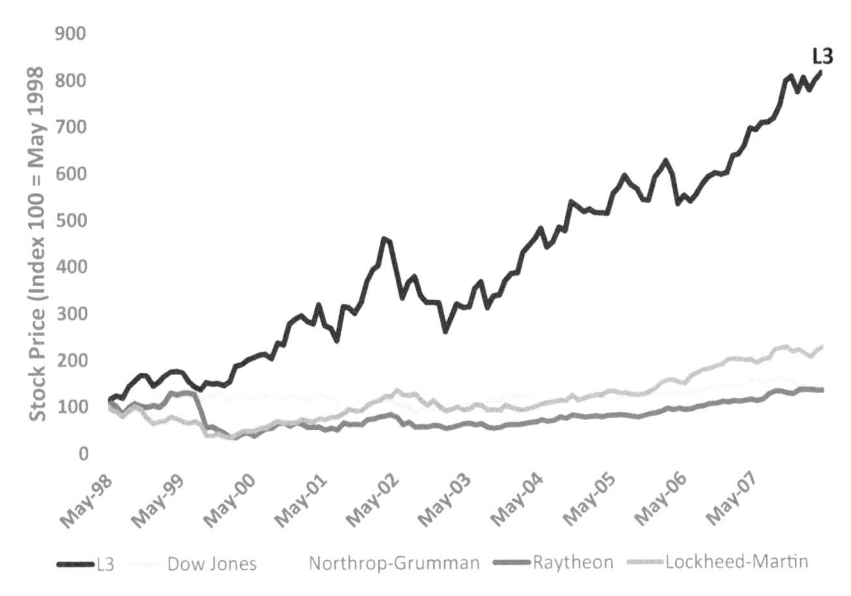

Source: Yahoo! Finance

2008-2018: L3 STOCK PRICE VS. DJI INDEX AND PEERS

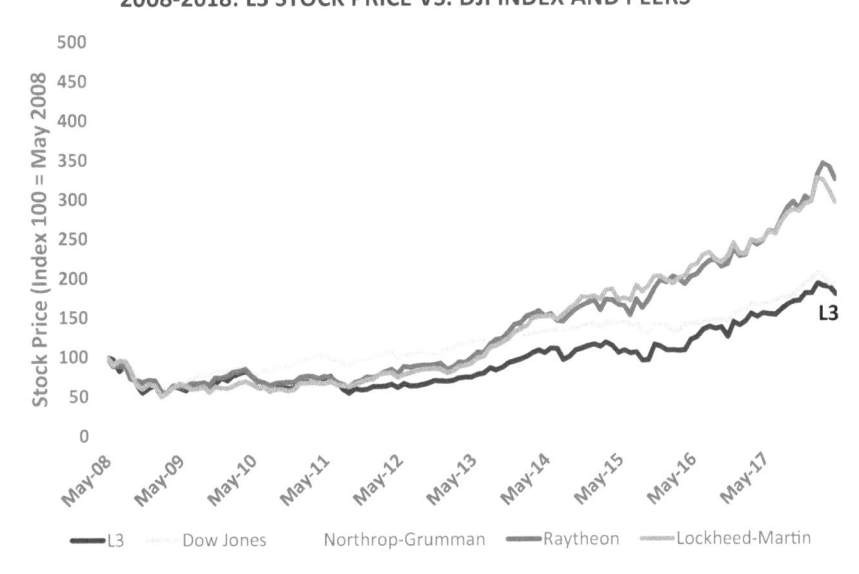

Source: Yahoo! Finance

KEY CONCEPT

VERTICAL VS. HORIZONTAL INTEGRATION

The A&D value chain is complex, with multiple levels of product integration, from components to modules to equipment and sub-systems and ultimately to complete systems, as illustrated in the following chart:

GENERAL STRUCTURE OF THE A&D VALUE CHAIN

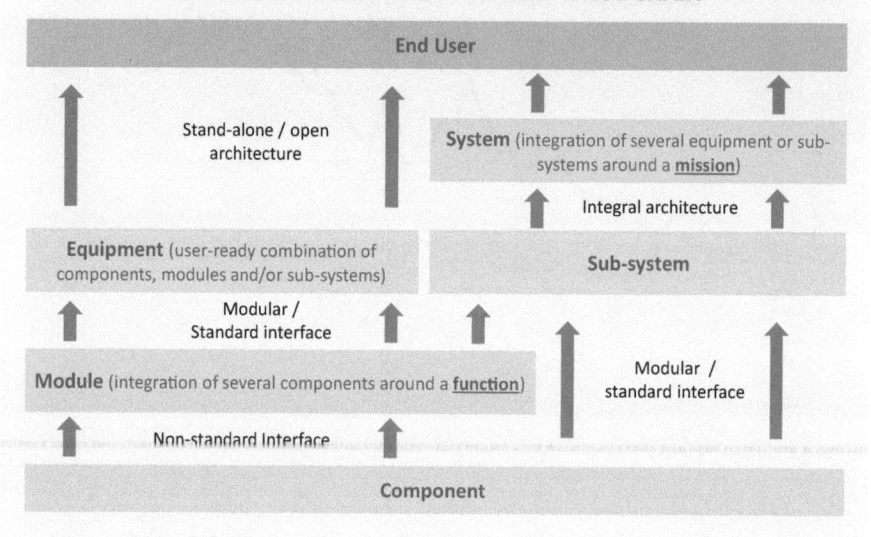

- End User
- Stand-alone / open architecture
- System (integration of several equipment or sub-systems around a **mission**)
- Integral architecture
- Equipment (user-ready combination of components, modules and/or sub-systems)
- Sub-system
- Modular / Standard interface
- Module (integration of several components around a **function**)
- Modular / standard interface
- Non-standard Interface
- Component

Source: Author

Historically, most players in the industry have favored vertical integration as a way to enhance their added value. Indeed, it is natural for - say a component supplier to be eager to move up the value chain and offer a functional module or sub-system made up of several components. By doing so, they can hope to capture a bigger share of the overall value of an end-product.

But this is only the case in specific conditions. When there is value in combining several components to enhance product performance on certain dimensions valued by end user (e.g. functionality, operability, affordability . . .) and the interfacing of these components is non-standard, then it is worth developing a module around these components by integrating vertically (as opposed to just selling components to someone else who will do the integration).

Similarly, when changing a component technology or combining several modules enhances the performance of an equipment but disrupts the equipment architecture, then it is worth integrating vertically to propose a new equipment architecture that can accommodate the new component technology or the newly combined modules.

Conversely, if a component can be integrated directly into an equipment by using standard interfaces, then there is limited value in integrating vertically.

In a nutshell, moving up the value chain makes sense when there is value (i.e., in a way that enhances performance on dimensions valued by the end user) in combining different components, modules, or sub-systems via non-standard interfaces.

But L3, TransDigm and the likes (re. "The New Predators" column) have chosen a different model of value creation. Instead of moving up the value chain, they have expanded their portfolio horizontally at the level of functional modules and sub-systems (and associated services) and offered them as 'plug-and-play' boxes to their customers (typically the large system integrators), who can then integrate them in their own systems.

By not being tied up to a given product architecture, they are able to offer the same piece of equipment to multiple customers (thanks to its modularity), thus increasing their accessible market and bid-to-win ratio. Additionally, they are able to reap economies of horizontal scope (by replicating a similar value proposition and operating model across the portfolio) as well as some economies of scale (in corporate overhead functions: finance, high-level marketing, lobbying…).

Their customers, on the other hand, can focus on higher-level, mission-centric integration without having to worry about lower levels of functional integration.

Airbus at 50

March 2019

AIRBUS IS MARKING ITS 50TH anniversary this year and, coupled with the transition to a new management team, that makes this a perfect time to reflect on the state of the group.

There are plenty of reasons to celebrate, not the least the success in federating European countries behind a project that remains a benchmark for international cooperation and industrial integration—particularly among countries that were at war with each other just a quarter-century earlier.

But this is also an opportunity to wonder if the noted achievements are truly worthy of celebration. Has Airbus really lived up to its potential? Has the company's management performance been so impressive to deserve a top mark? Is the group healthy enough today to guarantee that it will still be around 50 years from now? The honest answers must be "no."

The group's profitability has remained desperately poor and the business portfolio dangerously imbalanced.

Surely, one cannot laud the group's success today because it has managed to catch up and compete head-to-head with Boeing in the commercial aircraft business. This was achieved more than 15 years ago! Since then, CEOs have come and gone, strategic plans have been published, improvement programs and innovation initiatives have been implemented, even the company's name has been changed, but the group's profitability has remained desperately poor and the business portfolio dangerously imbalanced.

One cannot look at the group's results without getting a sense of waste, partially captured by the one-off charges that keep showing up in the accounts to the extent that one may wonder if they are truly "one-off." Between 2014 and 2018, €7 billion ($8 billion) worth of such charges have been accounted for (out of €22 billion of cumulated profits), mainly related to the A400M program but also to the A350 and A380, in addition to predelivery payment mismatch, restructuring and compliance costs. Their impact has been softened only by some proceeds from asset disposals, primarily a financial stake in Dassault Aviation, along with the transfer of space assets to a joint venture with Safran and the sale of Defense Electronics. None of these disposals did anything to improve the underlying profitability of the group, however, which is stuck at around 5%, far from the 10% management target (*see chart below*).

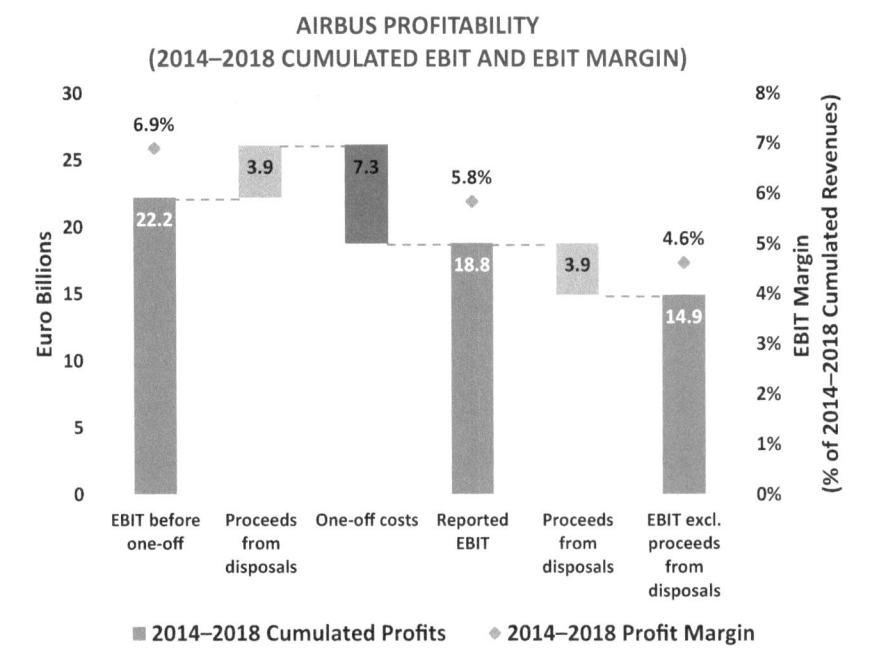

AIRBUS PROFITABILITY
(2014–2018 CUMULATED EBIT AND EBIT MARGIN)

	EBIT before one-off	Proceeds from disposals	One-off costs	Reported EBIT	Proceeds from disposals	EBIT excl. proceeds from disposals
2014–2018 Cumulated Profits (Euro Billions)	22.2	3.9	7.3	18.8	3.9	14.9
2014–2018 Profit Margin	6.9%			5.8%		4.6%

Source: Airbus Annual Reports, Author

The business portfolio is even more concerning. Airbus today is essentially one (arguably huge) cash cow and plenty of lame ducks: The cash cow is the A320 family of single-aisle commercial airplanes, which Airbus factories are churning out at the rate of 58 per month, expected to go up to 65 or more within the next couple of years. That is 80% of all commercial airplanes delivered annually and close to 50% of the group's total revenues (and probably at least three-quarters of its profits). Beyond that, the A350 production forecasts keep being revised downward, the A330 is at a standstill, the A380 has just been cancelled, and the A400M is a financial gulf.

The other divisions are not making the picture any brighter: The helicopter business has been struggling with decreasing revenues and poor profitability for

> *Airbus today is essentially one (arguably huge) cash cow and plenty of lame ducks.*

several years because of difficult market conditions, aging products, and operational mishaps. The defense business always has been and still is in a strategic no man's land, being a collection of legacy national businesses, politically driven export contracts and subscale security and service activities. Finally, the space sector's two flagship programs—the Ariane 6 launcher and OneWeb satellites— meant to be Airbus' answer to New Space, are close to being stillborn due to lack of competitiveness for the former and huge cost overruns for the latter*.

Adding to this grim landscape, the group is under investigation by British and French authorities for allegations of fraud, bribery and corruption. This could not only lead to significant financial penalties (including by the US authorities), it also could weaken the group's foundations as, in this process, Airbus has given access to highly sensitive information to outsiders (most noticeably American law firms) and instilled a climate of distrust, if not despair, within the company**.

So, while 2019 should be a year to celebrate and take stock of the achievements of the last half-century, it must also be a wake-up call for Airbus. The challenges ahead are huge, and the new management team's response could well decide whether the company will still be around in 50 years. At the top of the new CEO's to-do list should be a complete revamping of management functions to finally get rid of the bureaucracy and politics that have plagued the group since its onset.

UPDATE:

OneWeb filed for bankruptcy in March 2020 and was taken over in July 2020 by a new team of investors made up of the British Government and Indian conglomerate Bharti.

**In January 2020, Airbus announced that it had reached a €3.6bn-settlement with French, British and US authorities.*

KEY CONCEPT

PRODUCT PORTFOLIO ASSESSMENT

The most used framework to assess a product portfolio is the so-called BCG matrix, that classifies products or product lines along two dimensions (market growth rate and market share), and into four categories: Stars, Question Marks, Cash Cows and Dogs (or Lame Ducks). Each category calls for a different set of strategies to either grow, protect, improve, or turn around each product line.

A variant of the BCG Matrix classifies product lines along the two dimensions of top line (revenue) growth and profitability. In the case of Airbus, it would probably look somewhat like this (very roughly):

AIRBUS PRODUCT LINE PORTFOLIO

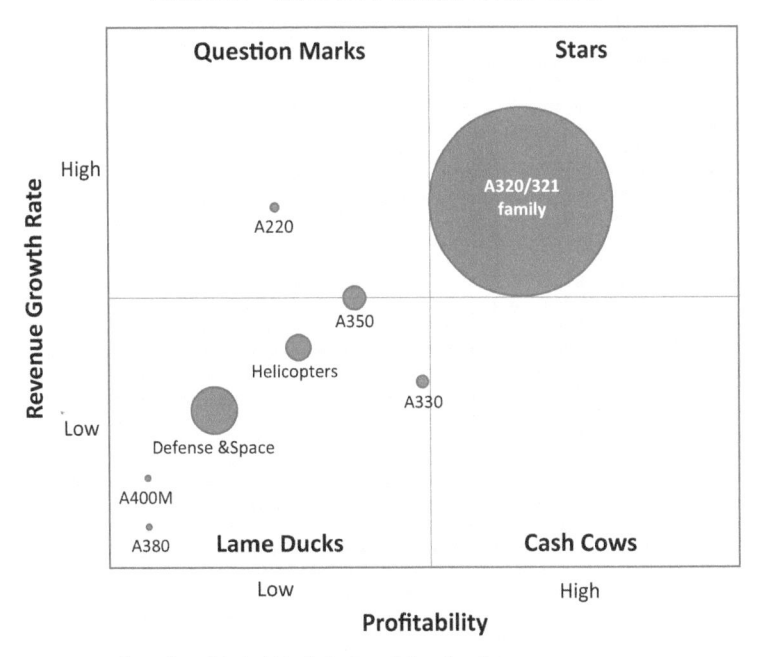

Note: Size of the bubbles indicates relative size of revenues

Hence the column's comment that Airbus today is essentially one (arguably huge) 'cash cow' (more of a 'star' actually) and plenty of 'lame ducks'. It is also important to note that "Defense & Space" and "Helicopters" are themselves made up of different product lines, so they should be broken down further to do the assessment properly. For example, it is to be expected that within "Defense & Space", the missile (via MBDA JV) and military aircraft activities (via Eurofighter JV) are relatively profitable, albeit with limited top line growth, and so would qualify as 'cash cows' in the overall Airbus portfolio.

Doomed Strategy (Bombardier)

August 2019

WITH THE SALE OF ITS CRJ BUSINESS to Mitsubishi announced in June, following the transfer of the C Series' majority ownership to Airbus last year, Bombardier essentially exits a sector it entered 33 years ago when it acquired Canadair from the Canadian government. In the meantime, close to 2,000 CRJ and 1,300 Q-Series airplanes have been produced and delivered, and a brand-new aircraft—the C Series—has been developed, but the tenure of Bombardier in commercial aerospace will ultimately be remembered as a financial and strategic failure.

It all started with a strategic dilemma at the end of the 1990s. Bombardier had been remarkably successful with its CRJ family of airplanes, stretching its original 50-seat version into 70- and 86-seat ones, but it was reaching the end of what could be done with this generation of aircraft and within the regional segment. Bombardier executives realized that while the 100-150-seat segment was already occupied by the Airbus A318/319 and Boeing 737-600/-700, these products were not optimized for that segment. This opened an opportunity for Bombardier to design an aircraft that would specifically address the segment's needs and thus compete advantageously with the incumbents.

After several false starts, the urge to design a new aircraft proved too strong to resist, and a revamped C Series program was launched in 2008 with an ambitious road map. The aim was to capture 50% of a market (100-150 seats) estimated at 6,000+ airplanes over 20 years and offer "best-in-class" performance for fuel efficiency, cash operating costs and CO_2 emissions, thanks to major innovations such as resin infusion and geared turbofan technologies. It was a daring project, to say the least.

The C Series was not just about a product, as good as it may be, competing with another, but rather about a relatively small Canadian company facing off with two global giants in a market they had "duopolized" for decades.

And it turned out to be too big a challenge for the company: After spending around $5.5 billion to develop the aircraft (more than twice as much as initially planned) and losing billions in the process, Bombardier's leaders realized that carrying it on would probably bring the whole company down. With the C Series' future hanging by a thread and its two other commercial aircraft programs (CRJ and Q-Series) reaching the end of their lives, it chose to exit completely.

In hindsight, Bombardier's management made several strategic mistakes along the way that could have been avoided, had they been more lucid about their market and the industry dynamics at play.

The first mistake was not realizing that the segment they were targeting with the C Series was not just an extension of the markets they knew. It was a completely different market, with different requirements, different success factors and two incumbents with a huge competitive advantage in terms of the installed base, marketing clout and economies of scale in design, production and customer support. In essence, the C Series was not just about a product, as good as it may be, competing with another, but rather about a relatively small Canadian company facing off with two global giants in a market they had "duopolized" for decades.

The second mistake was believing Bombardier's experience in developing smaller aircraft (28 in 20 years!) would give it a head start and take it up the learning curve quickly for the C Series as well. But with so many innovations involved and such a short timeframe to deliver (five years between the official launch and entry into service), it was wishful thinking, and Bombardier had to relearn pretty much everything from scratch about designing and managing a new aircraft program.

The third mistake was underestimating the challenge of getting its supply chain to deliver on specifications, on time and on cost. With some daring outsourcing decisions (noticeably the one to have part of the fuselage made in China) and without proper supply chain management processes in place, Bombardier quickly found itself at the mercy of its suppliers, which ultimately drove costs out of control. It is no surprise that Airbus is revisiting its supply chain to make the A220 program profitable.

All in all, Bombardier management failed at the most basic step of corporate decision-making: analyzing industry dynamics and using that analysis to inform its strategic decisions. Instead of going after a market where entry barriers were huge,

> *Bombardier management failed at the most basic step of corporate decision-making: analyzing industry dynamics and using that analysis to inform its strategic decisions.*

where the company's bargaining power (with buyers and suppliers) was low and where rivalry was already intense,

Bombardier would have been better off sticking to its core business and protecting its market share from new entrants such as Mitsubishi, Comac and Sukhoi while keeping the pressure on Embraer and ATR—unless it expected from the beginning to sell the C Series program to Airbus or Boeing. But in that case, Bombardier executives probably would have had different number in mind than the symbolic $1 Airbus ended up paying for the takeover.

KEY CONCEPT

MARKET ATTRACTIVENESS

A basic analysis of the attractiveness of the large passenger aircraft market (using Michael Porter's classic 'Five Forces' framework), shows how risky this business should have appeared to Bombardier from the get-go:

- **Competitors' Rivalry: High** – The competitive rivalry between Airbus and Boeing was already intense and it was foreseeable they would not let Bombardier come close to their core business.
- **Customers' Bargaining Power: High** – Airline and leasing customers have a lot of bargaining power, and have consistently played Airbus and Boeing against each other, leading to huge price discounts. It was expected that they would be at least as aggressive with a new entrant like Bombardier, who needed to build an order book from scratch.
- **Suppliers' Bargaining Power: Medium** – Suppliers tend to have less bargaining power than customers. However, Bombardier C-Series potential volume was small compared to the volumes offered by Airbus and Boeing, which made suppliers' position stronger than usual. In other words, Bombardier needed them more than they needed Bombardier.
- **Entry Barriers: High** – Entry barriers are extremely high, due to the fixed costs and capital expenditures involved in designing and making a large passenger airplane. Bombardier not only underestimated them, but the financial commitment was disproportionate compared to the company's size (and to the Canadian Government's geopolitical clout).
- **Risk of Substitution: Low** – The risk of substitution is low (except on short-haul, inland routes, where the train is an obvious substitute), which is an advantage for incumbents, but is irrelevant to a new entrant with a non-disruptive product.

FALLEN ICON
(ROLLS-ROYCE)

OCTOBER 2020

LAST WEEK, ROLLS-ROYCE ANNOUNCED it was evaluating a cash call of up to £2.5bn ($3bn) to help shore up its balance sheet as it keeps bleeding money amid the worst crisis in its history. This is just the latest piece of – bad – news about a company whose fate looks increasingly like the one of a fallen hero in a Greek tragedy: a victim of both misfortune and its own hamartia, punished by the gods and forced to navigate through titanic storms and raging seas, until it either finds redemption or suffers eternal damnation.

There is indeed much to feel sorry for Rolls-Royce, once the jewel of the British high-tech industry. After suffering a record $6.75bn loss for the first half of the year and witnessing its credit rating being downgraded to junk by Standard & Poor's in May, the recent cash call announcement has now brought its stock price to the ground (*see chart below*). The 100-year-old, $19bn-revenue corporation, which once stood for engineering artistry and world-class technology, is now valued at less than $4bn. By comparison, 17-year-old Tesla, which is about the same size in revenues and staff, is valued at close to $400bn!

Such a reversal of fortune raises a lot of questions about what led to such tragedy. Of course, one cannot blame the company for finding itself in the midst of an unprecedented collapse in air travel. But one could argue that this crisis is just laying bare and amplifying Rolls-Royce's underlying weaknesses and past mistakes. Ultimately, a company's success rests on three main pillars: culture, strategy, and leadership. It looks like Rolls-Royce lost its way on each of these.

> *Ultimately, a company's success rests on three main pillars: culture, strategy, and leadership. It looks like Rolls-Royce lost its way on each of these.*

Culture shapes a company's resilience, its ability to consistently perform under any circumstances. Rolls-Royce's culture of innovation, superior quality and work ethics used to be its strongest asset. Yet, over the last few years, lingering quality issues with the Trent 1000 engine have cost the company billions in corrective actions and undermined its reputation with customers. This reputation was further damaged when Rolls-Royce was indicted in the

UK and the US on multiple counts of conspiracy to corrupt, false accounting and failure to prevent bribery over the span of three decades. The case was settled in 2017 with the payment of a $800m fine. Somehow, over time, Rolls-Royce's business culture had strayed far away from what its legendary Chief Executive Ernest Hives once described as the "moral business standard of the Rolls-Royce company", adding: "We must never allow a question of profit to jeopardize this position".

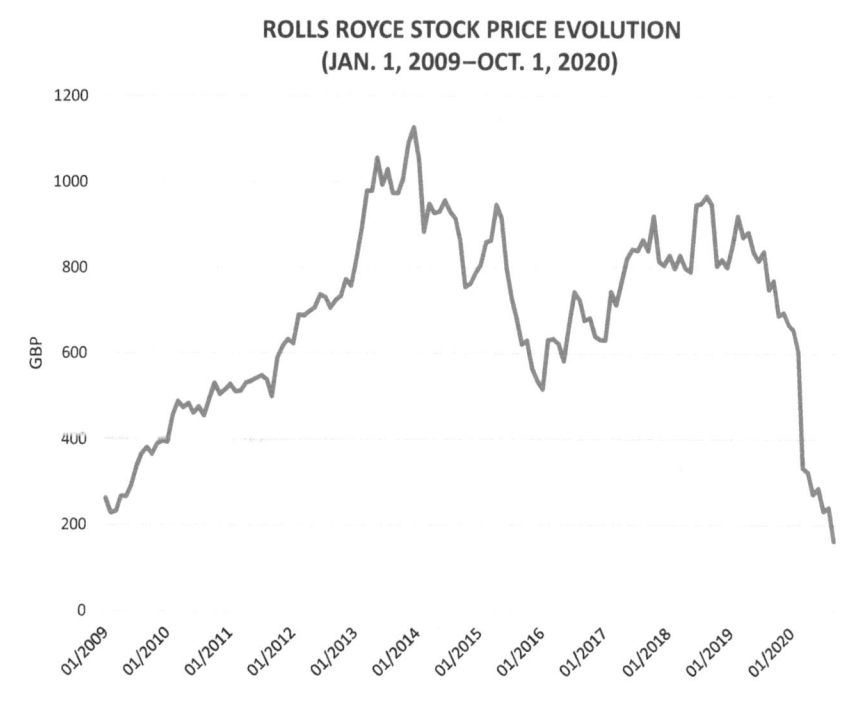

ROLLS ROYCE STOCK PRICE EVOLUTION (JAN. 1, 2009–OCT. 1, 2020)

Source: Yahoo! Finance

While product quality and commercial practices are somewhat unrelated, it does suggest that the company's culture has been impacted by an increasing focus on financial management and shareholders' return. It is therefore probably not a complete coincidence if the company's stock price increased fivefold between 2009 and 2013 while problems were building up below the surface.

As far as strategy is concerned, the most questionable decision was the company's exit from the lucrative single-aisle market. While Rolls-Royce was a founding member of the IAE consortium that developed the V2500 turbofan for the Airbus A320 family, it sold its stake to Pratt & Whitney (P&W) in 2012. By doing so, Rolls-Royce not only exited the most profitable aeroengine market segment but also marginalized itself at a time where its main competitors – GE, Safran and P&W owner UTC – were all beefing up their product and service portfolio to positioned themselves as "super tier-ones".

Ironically, it was the second time in its history that P&W – as the big beneficiary of this exit – could thank Rolls-Royce for its luck. Already, at the beginning of the Cold War, it was the experience that Pratt gained from a licensing contract to produce two Rolls Royce military engines that had enabled it to become one of Rolls-Royce's major competitors.

As for the company's leadership, it is hard not to notice how many top executives have come and gone over the last decade. Since 2010, there have been three CEOs, four CFOs and four different heads of the civil aerospace business. Such a high management turnover, unprecedented in Rolls-Royce history, certainly impacted the company's performance and is either a cause or a consequence of a failed leadership.

> *Its culture, strategy and leadership have all but reinforced an insularity that, compounded with Britain's imminent exit from the European Union, could prove fatal.*

At least one thing has not changed at Rolls-Royce: its Britishness. Almost all top executives and board members have always been and still are British. In a global industry like aerospace, this is maybe the firm's most tragic flaw: its culture, strategy and leadership have all but reinforced an insularity that, compounded with Britain's imminent exit from the European Union, could prove fatal.

Part Three

INNOVATION MODELS AND... THE BIG(GER) PICTURE

Chapter 7

STRATEGIC INNOVATION

"The finger pointing to the Moon is not the Moon."

Buddhist Saying

CHAPTER INTRODUCTION

This chapter is about innovation in the A&D industry.

The first column discusses the concept of disruptive innovation, a theory developed by academic and business thought leader Clay Christensen that became widely popular in the last two decades, including in the aerospace industry. Despite its long history of government-protected markets, the A&D industry has not been immune to this phenomenon and more disruptions are on the way.

As a potential domain of application of this theory, the second column looks at what it would take to disrupt the large commercial aircraft market currently monopolized by Boeing and Airbus. It would certainly be feasible technically and financially, but more ingredients would be needed to make it happen.

The next two columns discuss two features of the industry that have become obstacles to innovation: The conservative behavior of customers, particularly in defense, and the reliance on institutional customers to fund R&D projects.

The chapter's last column is an assessment of Airbus Group' recent attempt at revamping its entire innovation strategy in a dramatic – possibly traumatic – way, highlighting both the rationale and the risks of such endeavor. While such innovation drive may be the only way to keep Airbus relevant in the digital economy, it also carries the risk of damaging the DNA of a company whose products stand for reliability and safety.

DISRUPTIVE INNOVATIONS
(REMEMBERING CLAY CHRISTENSEN)

FEBRUARY 2020

CLAY CHRISTENSEN, AN OUTSTANDING academic and business thought leader, passed away last month. He was the father of the disruptive innovation theory that became widely popular in the last two decades, including in the aerospace industry. I was lucky enough to have him as a professor at Harvard Business School at the time when he was still fine-tuning his theory and before he became a best-selling author with his first book, The Innovator's Dilemma. Christensen's mind was as brilliant as his heart was compassionate. His fundamental premise was that innovation could become a much more predictable and manageable science if business leaders would only focus on the right parameters, such as the resource allocation process, technology performance trajectories and value chain dynamics of modularity and integration.

Innovation could become a much more predictable and manageable science if business leaders would only focus on the right parameters.

The main insight from his theory was that incumbents in any industry tend to end up overshooting what their customers really need. By doing so, they open the door to new entrants that disturb the way traditional players have been making money by offering a typically less sophisticated technology, but one better fitted to the needs of less demanding customers. The aerospace and defense industry, despite its long history of government-protected markets, has not been immune to this phenomenon.

The most obvious example is the way 'low-cost' (or low fare) airlines have redefined the economics of air travel, by focusing on basic functionality and drastically reducing the cost of air travel, thus allowing a whole new category of consumers to fly. As with any disruptive innovation, it was enabled by an "infrastructure" innovation, which in this instance was the combination of airline deregulation and of an aircraft product (twinjet narrow-body) perfectly fitted for short-haul trips.

Another example of disruptive innovation according to Christensen's theory was the Joint Direct Attack Munition or JDAM—built by Boeing for the US military—that became particularly successful in the late 1990s. The

JDAM was essentially a "dumb" bomb made smarter with the addition of a set of maneuverable fins and a GPS guidance system. While initially designed to address low-value targets at short range, it became the weapon of choice for most air-to-ground missions in Afghanistan at a fraction of the cost of missiles such as the Tomahawk, the capabilities of which "overshot" the needs of the military for most missions. A similar thing happened in Europe with the AASM guided bomb superseding missiles for most Rafale air-to-ground missions over the last 20 years. Like JDAM, the AASM turned out to be "good enough" for a large majority of military customers' needs, thus disrupting traditional missile manufacturers.

In the space sector, Elon Musk obviously disrupted the launch sector but, interestingly, the disruptive part of it was not that he addressed a low-end segment with a less capable technology. In fact, from the beginning, Musk targeted the high-end segment of the market (expensive and large geosynchronous [GEO] government and commercial satellites) with a fairly traditional technology. What was truly disruptive about SpaceX initially was that Musk created his own ecosystem, designing his own rocket, vertically integrating its production and acquiring his own test range, which was something straight out of Christensen's book. As a new entrant, it is indeed extremely hard to disrupt an industry without creating a whole new ecosystem of your own, simply because there are too many vested interests in the existing one. In Christensen's lingo, you are more likely to succeed as a disruptor with an "architectural" innovation than with a "component technology" innovation.

Another recent disruptive innovation in space has been the emergence of CubeSats. Not so long ago, the trend was for satellites to become ever bigger, more complex and more expensive. With CubeSat (shoebox-size) technology, the space industry finds itself on a radically different performance trajectory (*see graph below*). As with most disruptive innovations, it started by addressing the needs of "low-end" customers (universities) with a basic technology, but performance is continuously improving to the point where it is starting to compete with traditional satellites for Earth-observation and telecommunication missions.

> *As a new entrant, you are more likely to succeed as a disruptor with an "architectural" innovation than with a "component technology" innovation.*

So, what will be the next disruptive aerospace innovations, according to Christensen's theory? Most probably drones and unmanned aircraft for the military aircraft market, and possibly—further down the line—flying taxis or private flying vehicles for commercial aviation. Indeed, a key characteristic of a disruptive innovation is that it enables unskilled people to do things that only experts could do before. This will clearly be the case if, one day, getting a pilot's license for a flying car—autonomous or not—is as easy as getting a driver's license today, or supersedes it altogether because our urban transportation systems will have become fully three dimensional. That will be a disruption, all right.

> **KEY CONCEPT**
>
> ### THE DISRUPTIVE INNOVATION MODEL
>
> The disruptive innovation model relates to the concept of **performance trajectories**—the rate at which the performance of a product or service improves over time. There are two types of performance trajectories in every market. First, there is a trajectory measuring **the improvement in a product or service that customers can absorb or utilize over time**, depicted by the three parallel dotted lines in the diagram below. The top line represents the most demanding customer tier (in this case, military), the bottom line the least demanding customer tier (in this case, academia).
>
> The second trajectory, shown by the two solid lines in the diagram, suggests that there is a distinctly different trajectory of improvement that the innovators in the industry provide, as they introduce new and improved products. Usually, this second trajectory—**the pace of technological innovation**—outstrips the ability of customers in a given tier of the market to absorb it. This means that a company whose products and services are designed for what customers in the mainstream market need today will usually overshoot what those same customers can actually use tomorrow. They do so because, by selling more sophisticated products to more demanding customers at the high end, companies typically can improve their profit margins.
>
> This propensity of incumbents to overshoot the performance that customers can absorb creates an opportunity for innovative companies with "disruptive technologies"—cheaper, simpler, more convenient products or services (such as CubeSat technology)—to enter the tiers of the market in which customers are most overserved by the prevailing offerings (e.g., universities).
>
> Almost always, the leading companies in industries where this happens are so absorbed with sustaining innovations—the upmarket innovations

that enable them to address more sophisticated and profitable customers in the more demading tiers of the market—that they miss (or dismiss) the disruptive innovations piercing into the market from the low end.

When compared to the mainstream product, disruptive technologies are inevitably inferior in terms of performance. Often though, they enable a larger population of less-skilled or less-wealthy individuals to do things once reserved for specialists. Which is exactly the case with CubeSat technology.

Ultimately, what tends to happen is that the new technology's performance will keep improving (shown by the solid line's dotted extension in the diagram) to the point where it will become "good enough" for the needs of mainstream customer tiers. At that point, the new technology will become truly disruptive as new entrants will start taking market shares from traditional players and impairing the way they have been making money.

THE DISRUPTIVE INNOVATION MODEL APPLIED TO SATELLITE TECHNOLOGY

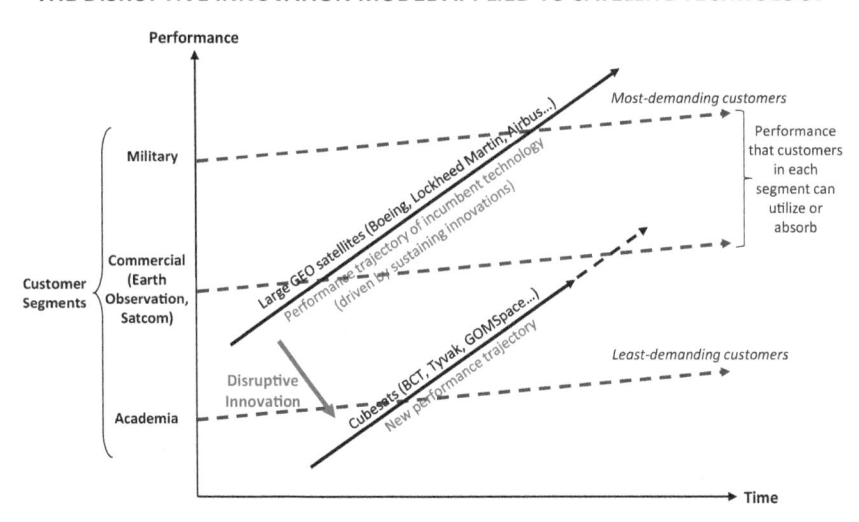

Source: Clay Christensen's Disruptive Innovation Theory, Author

Creative Destruction

March 2016

AS SPACEX, BLUE ORIGIN AND other space disruptors are on their way to making expendable launchers and $200 million satellite launches a thing of the past, one may wonder what the future holds for large commercial aircraft manufacturers Boeing and Airbus. Should they start worrying? How long will it take for someone to do to airplanes what SpaceX did to launchers?

Indeed, as the waves of the digital tsunami keep crashing onto the shores of the Old Economy, industries are falling off the cliff one after the other: Telecommunications, media, banking, retail . . . Automotive is clearly next, with the advent of driverless cars and innovative mobility solutions that are likely to revolutionize ground transportation. Yet, amid this landscape of creative destruction, one industry seems to hold out against the invaders: commercial aerospace.

Looking at the way an airplane is made today, it looks incredibly complex and old-fashioned compared to other high-technology products: Millions of parts, hundreds of miles of electrical wires, tons of machined and formed metal, thousands of suppliers making bits and pieces, spread out all over the globe. It suffices to look at the wing of a commercial jet aircraft during landing and notice the intricate network of pipes and cables that run through to get a glimpse of that complexity.

On the face of it, therefore, a large commercial aircraft is a very expensive product to build, with a steep learning curve and huge entry barriers. The Chinese are getting there slowly, but only by replicating what Airbus and Boeing are doing, and so they are unlikely to find themselves on a very different "cost curve."

Today's aircraft architecture is the result of years of incremental evolutions that have added complexity, rigidity and cost to the manufacturing process and the product itself and has hence continuously pushed prices upward.

But could there be another, much more disruptive, way to break through? Today's aircraft architecture is the result of years of incremental evolutions that have added complexity, rigidity, and cost to the manufacturing process and the product itself and has hence continuously pushed prices upward. It is reasonable to assume that anybody starting with a clean sheet could do things very differently and so cut the cost of making an airplane at least by half. Three-dimensional printing is meant to change the economics of aircraft manufacturing; wireless connectivity and innovative materials can do wonders to make an aircraft lighter and simpler; and a more compact and integrated supply chain would also save significant time and money.

Assuming it is feasible from a technical standpoint, would it be financially and commercially viable? Could a new player realistically break the duopoly of Airbus and Boeing with a radically different approach? To answer that question, one needs to consider the four ingredients of a successful venture: money, talent, business model and vision.

Money is not the issue. The recent accumulation of wealth by a small number of individuals and companies on the planet makes it plausible for one of them to fork out a few billion dollars in seed money for such endeavor. A few percent of Apple's or Google's treasure chest would do the trick. Furthermore, venture capitalists are looking for ever bigger bets in a world that is increasingly unpredictable and seems to have no upper limit for potential returns on investment. They would be more than happy to get involved.

Talent should not be an issue either. SpaceX and other "New Space" ventures have demonstrated that thousands of talented engineers would love to leave the bureaucracy and stuffy atmosphere of incumbent companies for the excitement of something truly new and visionary.

As for the business model, it would have to rely on a clear value proposition and well-thought-out market entry strategy, but with the right resources and processes, there is no reason why what the Japanese have achieved in the automotive industry in the 1980s could not be achieved by someone in the aerospace industry today. What will be needed, however, is a "system" approach, meaning that it should not only be about producing cheaper airplanes, but also about creating a whole new ecosystem around designing, building and possibly operating them.

And that leads us to the final ingredient: vision. The goal of simply making a cheaper airplane is not aspirational enough. To seduce investors, engineers, airlines and passengers, the vision must embrace something that stretches the imagination,

To seduce investors, engineers, airlines and passengers, the vision must embrace something that stretches the imagination, something that could revolutionize the experience and economics of air travel altogether.

something that could revolutionize the experience and economics of air travel altogether, with a level of ambition embodied by the aviation pioneers of the early twentieth century. And that is probably the biggest hurdle of all: Nowadays, those visionary, charismatic and daring individuals are a rare commodity.

> **KEY CONCEPT**
>
> ### Ecosystem
>
> A business ecosystem consists of three elements, similar to those of a biological ecosystem:
>
> 1. A **value network**: a network of interactions towards a shared purpose (e.g., delivering 'value' in the shape of a product or service to customers).
> 2. A **"physical" environment**: a common infrastructure or technological paradigm (acting as an enabler).
> 3. A **community**: a set of "stakeholders" that are linked together and interact through the shared purpose and the common infrastructure.
>
> Often, an ecosystem takes shape around a dominant player or group of players, thus creating significant barriers to entry (e.g., Apple, Microsoft, The Automotive Big Three (in the 1970's), Boeing and Airbus duopoly…). This implies that for a new entrant trying to break through, it may be necessary to create a separate ecosystem.
>
> This is how Toyota broke through the US automotive market in the 1980's. It created an entirely separate set of upstream (suppliers) and downstream (distribution and retail outlets) partners. In this case, the Toyota ecosystem relied on the Toyota Production System as a key enabling infrastructure to create a community of stakeholders linked together and driven by a shared purpose.

COMFORT ZONE

AUGUST 2014

IT IS A STRIKING FEATURE OF the aerospace & defense (A&D) industry that, by and large, it is still dominated by the same companies as thirty years ago while in so many other industries, historical players have long been challenged and sometimes defeated: General Motors, Digital Equipment, Xerox, Motorola, Texas Instruments and Eastman Kodak, to name of few Fortune 500 companies of the 1990s that have faced major challenges during the last two decades.

Yet, in A&D, apart from some reshuffling and rebranding, the top players are still the same: Boeing, Lockheed (plus Martin), Northrop (with Grumman), Raytheon, United Technologies, British Aerospace (now BAE), Thomson-CSF (now Thales), Snecma (now Safran), Aerospatiale and DASA (now Airbus Group). Out of the world's top 50 A&D players—putting aside straight spinoffs such as Alliant Techsystems, Spirit AeroSystems, Exelis, Huntington Ingalls or Leidos—only two companies were founded within the last three decades: B/E Aerospace (1987) in commercial aerospace; and L-3 Communications (1997) in the defense sector (while L-3 was the result of spinoffs from Lockheed Martin, its business model was so innovative that it could have been considered a new company).

There are two primary reasons for such stability: the relative youth of the commercial aerospace sector and the conservative behavior of customers, particularly in defense.

In a relatively young and immature sector such as commercial aerospace—the market only really took off in the 1980s—customers have logically placed a premium on product reliability as the primary dimension of performance, over other elements such as convenience or price. This has favored companies that control the design and final integration of their products: primarily Airbus and Boeing, as well as their major suppliers, which themselves created "closed systems" around key aircraft subsystems such as engines, avionics, power systems and control systems. All these

There are two primary reasons why the A&D industry is, by and large, still dominated by the same companies as thirty years ago: the relative youth of the commercial aerospace sector and the conservative behavior of customers, particularly in defense.

dependencies have made it nearly impossible for new players to break in with innovations, at least in the "design and integration" markets dominated by incumbent players.

The only areas where new players like B/E Aerospace have prospered have been cabin equipment and consumables (as well as some MRO services), sectors where convenience, price, and the ability to customize have overtaken reliability as the key performance dimensions for airlines. In such a context, opportunities are created for new players to come in with different value propositions, for example as "aggregators" (one-stop-shops) or brokers of products and services contributing specific functions or filling gaps in the industry value chain.

However, until visionary and deep-pocketed players of the same breed as Elon Musk or Jeff Bezos develop new concepts for air transport or aircraft design and manufacturing—such as SpaceX in the space sector today or Toyota in its time in the automotive sector—it will be quite difficult to challenge the industry leadership, including aircraft OEMs and their major suppliers.

Meanwhile, in defense, military customers' propensity to engage in ambitious and large programs in closed architectures has protected long-established players by placing a premium on product reliability and technological complexity, and therefore on design and integration skills. Over time, this has created a kind of comfort zone for both customers and large established players and has become barrier to entry for new players with different value propositions.

It took the creation of L-3 Communications in 1997 to show that another value proposition was possible, namely one that consisted in creating a "Sears catalog" of the defense sector (as co-founder Frank Lanza once put it), positioning the new company as a horizontal aggregator of products and services. This represented a complete break with the traditional "vertical integrator" model of the top defense primes. However, military customers are even more conservative than commercial airlines, so the L-3 story remains an exception. In fact, over the last few years, L-3 has changed its model back to a more traditional one—probably encouraged to do so by the US Defense Department's behavior.

While conservatism has brought us air travel safety and—for most of us—(homeland) security, the A&D industry as we know it in North America and Western Europe is living dangerously off antiquated assets, cozy relationships, and outdated practices. Its long-term prosperity will only be ensured if more diverse players—including non-A&D specialists and small and medium enterprises— are allowed to compete on a level playing field, and if customers are willing to drastically change the way they procure their products and services. Undergoing this required evolution, if not revolution, in customer behavior is not the least of the A&D industry challenges.

> *While conservatism has brought us air travel safety and—for most of us—(homeland) security, the A&D industry is living dangerously off antiquated assets, cozy relationships and outdated practices.*

KEY CONCEPT

VALUE PROPOSITION

Value proposition is a key ingredient of a company's innovation strategy. It drives several "hard factors" (business model, cost structure, channel strategy...) and therefore engages the company in a specific direction. It can be defined along two main dimensions:

- **'Product intimacy'**: the ability to leverage an expertise or know-how about, or a privileged or proprietary access to, a system, infrastructure, asset, or technology. The more complex the underlying technology, the more valuable product intimacy is.
- **'Customer intimacy'**: the ability to leverage an advantageous position with customers due to either a specific understanding of the customer's processes or mission, or a specific relationship that creates goodwill and trust. The more complex the underlying mission, the more valuable customer intimacy is.

Thus, four generic value propositions can be identified, each with a different primary source of value and differentiation:

- **'Innovator'**: High level of product intimacy, allowing to develop, produce and deliver innovative and proprietary products and/or services that can be offered to a wide range of customers.
- **'Integrator'**: Combination of a high level of product and of customer intimacy, allowing to offer customized and integrated solutions to meet the unique needs of selected customers.
- **'Aggregator'**: High level of customer intimacy, allowing to play the role of aggregator ('one-stop-shop') of products and/or services for selected customers, without necessarily relying on a high level of product intimacy.
- **'Broker'**: Ability to contribute specific functionalities or filling specific "value gaps" in a given value chain environment, without relying on a high level of product or customer intimacy (e.g., syndicator, distributor, API software...).

Different value propositions call for different business scopes, channel strategies, underlying capabilities, and delivery mechanisms. They also entail different types of risks, such as over-complexity or commoditization for value propositions that rely exclusively on a high level of product intimacy, or loss of customer's trust for those that rely on a privileged position (high customer intimacy) to offer « commodity » services at a premium.

As an illustration, in the following matrix, several Boeing's businesses have been positioned according to their (perceived) predominant value proposition:

GENERIC VALUE PROPOSITION FRAMEWORK
(illustrated with Boeing's businesses)

	Low Customer Intimacy	**High Customer Intimacy**
High Product Intimacy	**'Innovator'** Boeing Commercial Airplanes Boeing Commercial Satellites Insitu Jeppesen	**'Integrator'** Boeing Defense Boeing Space (ISS, X-37B, SLS, ULA)
Low Product Intimacy	**'Broker'** HorizonX Boeing Distribution Services Analytx	**'Aggregator'** Aviall

Source: Author

Even though a company (or business within a company) may entertain more than one value proposition, most will strongly relate to and benefit from one specific value proposition (unless it is managed as a portfolio of independent businesses).

While a clearly articulated value proposition is a strong driver of performance, conversely, a lack of clarity in the business' value proposition can lead to underperformance, either by creating some internal misalignment, for example in allocating resources (e.g., should we invest in product development or distribution channels?) or by being misread by customers, who may then shy away from the company's offering.

DISRUPTIVE IMPERATIVE

DECEMBER 2014

ATTENDING THE REINVENTING SPACE Conference in London last month, it dawned on me that one of the biggest obstacles to innovation in the aerospace and defense (A&D) industry is the reliance on institutional customers to fund R&D projects. Indeed, the most innovative ventures today are commercially driven. Be it Skybox Imaging and Planet Labs with their low-cost Earth-observation satellites, SpaceX with its launchers or AeroVironment with its drones, they are all coming with a radically different, commercially driven approach to innovation. It does not mean that institutional customers do not have a role to play, but they are not in the driver's seat anymore.

> *One of the biggest obstacles to innovation in the aerospace and defense industry is the reliance on institutional customers to fund R&D projects.*

The traditional way of financing innovation in the A&D industry has been to rely on institutional customers such as the Department of Defense and NASA in the US, and national defense and space agencies in Europe. This worked as long as the technologies used for A&D applications were unique and ahead of those used in commercial fields. It created a kind of closed innovation ecosystem, with its own organization (e.g., prime contractors and system integrators), processes (such as technology readiness levels[iv], concept and technology demonstration or developmental test and evaluation) and cycles (five, 10 or more years). As with any closed system though, rigidity and bureaucracy have crept in and the whole system has become largely inefficient. That is how you end up with a price tag of $200 million for a fighter aircraft, satellite or launcher.

Sadly, most established A&D companies are still configured to play the "old game" and keep spending a lot of their time and dollars lobbying their institutional customers for R&D funding, which in the best case comes late and with plenty of strings attached—such as a required level of performance that typically translates into a 50% cost increase for 10% better reliability.

[iv] **Technology Readiness Levels:** https://www.nasa.gov/directorates/heo/scan/engineering/technology/txt_accordion1.html

Meanwhile, it has been clear for a while that the world is changing, and the historical "closed innovation" approach is no longer appropriate. The pioneer in that respect is probably Flir Systems, the U.S infrared specialist founded in 1978 that early on adopted a distinctive "commercially developed military qualified" (CDQM) strategy. By self-financing its R&D, the company was able to leverage its product development across multiple customers, both commercial and military. This CDQM strategy has become Flir Systems' trademark and given it an edge in terms of innovation, development costs and scale, which has translated into a substantially higher profitability than for its peers. Flir was also an early defense-sector adopter of commercial-off- the-shelf (COTS) technology.

Today, while the use of COTS components in the A&D industry has become more common, it is still not the default way of thinking about developing new products. But for disruptive players such as Skybox Imaging or Planet Labs, COTS is at the core of their strategy.

In fact, they both pride themselves on using almost exclusively standard commercial technologies—from complementary metal-oxide semiconductor image (CMOS) sensors used in digital cameras to electronic design tools used for the PlayStation 4—and yet achieve a similar, if not better, level of performance than their incumbent competitors. In some ways, they look more like consumer electronics or cell phone manufacturers than aerospace companies.

This disruptive dimension also can be found in a management philosophy strongly influenced by Silicon Valley's software expertise and venture capital- driven culture, both of which combine into a powerful innovation engine. Silicon Valley's well-recognized software concepts, such as "minimum viable configuration," rapid iteration or agile development, have been readily adopted by these start-ups to build their hardware competitively.

These examples may be dismissed as irrelevant to the highly demanding markets served by the large established A&D companies because they will not give customers the level of performance they need. But that is an illusion. Once customers—even

> *Once customers—even the most demanding ones—realize they can do their job with 90% of their requirements at 10% of the cost, they won't hesitate long before jumping ship.*

the most demanding ones—realize they can do their job with 90% of their requirements at 10% of the cost, they won't hesitate long before jumping ship. And this kind of step-change is now made possible by the digital economy: directly, because digital convergence blurs the frontiers between industries; and indirectly, because wealth created in sectors such as the Internet and consumer electronics is finding its way into the A&D industry—financing disruptive innovation in unmanned systems, satellites, airships and launchers, among other things.

A&D industry players should therefore not tie their future to the ability of institutional customers to fund their R&D. Instead, they should take the initiative and approach their innovation agenda more like start-ups do in the Silicon Valley, by being long on vision and talent and short on time and budget. If they manage to do that, then commercially driven innovation will become an opportunity rather than a threat.

KEY CONCEPT

CLOSED VS. OPEN INNOVATION

In closed innovation, a company relies on its own internal Research & Development (R&D) resources, processes and capabilities to develop new technologies and products that will then be commercialized by the company, while protecting the underlying IP (Intellectual Property) through patents or other IP protection mechanisms. It is typically a very linear process with clear and rigid milestones (such as the Technology Readiness Levels – TRL – commonly used in the A&D industry).

In the open innovation model, a company proactively taps into external sources for new ideas and innovations, instead of trying to develop everything on its own. Additionally, it is less protective of its own ideas, sharing them more proactively, and sometimes commercializing them straight out of the R&D department (in parallel to using the same ideas to develop and commercialize its own products). In such model, IP production and ownership is not an end in itself nor the most important innovation driver. Instead, the innovation process focuses on the ability to leverage both proprietary and open technologies or ideas to bring new products to market quickly and profitably.

As such the open innovation process is less linear and possibly more erratic, with multiple potential entry (and exit) points for both internal and external ideas into the product development funnel and 'go-to-market' process.

AIRBUS'S INNOVATION GAMBLE

JUNE 2017

FOR THE PAST TWO YEARS, Airbus has been revamping its innovation strategy in an unprecedented way: Investing early-on in OneWeb microsatellite constellation project, launching A3, a "disruptive innovation" outpost in the Silicon Valley, creating a corporate venture arm and appointing a new Group CTO, with a clear mission to shake things up. All these initiatives have been driven personally by Airbus' CEO Tom Enders, making them highly visible and scrutinized both inside and outside the Group.

Is this the kind of bold strategic move that will make the company as relevant and innovative in the decades to come as it has been in the past? Or is it a lame attempt at making the company and its CEO look "cool" in the age of digital natives, celebrity entrepreneurs and "unicorns" (start-ups valued at more than $1 billion)?

Enders' fascination with American-style management and business has been known for years and, as a new breed of aerospace entrepreneurs epitomized by SpaceX CEO Elon Musk was taking the industry by storm in the US, there was a clear sense of frustration that he and Airbus would not be taken more seriously when it came to being part of this new aerospace ecosystem. Enders therefore had no choice but to be bold and send strong signals to show that he, too, could play in that league.

And so, he took a series of initiatives to that effect: First, in 2015, he launched Airbus Ventures – a $150m corporate venture fund – and opened A3, both located in the Silicon Valley; then he invested $150m in OneWeb, and, finally, last year appointed former Google and DARPA engineer Paul Eremenko, initially hired as A3's CEO, as the Group's CTO to revamp the whole R&D organization and culture.

This last initiative has been the most controversial. While launching Airbus Ventures and A3 as stand-alone, dedicated organizations were textbook-moves as to how a large organization should deal with disruptive innovations, handing over the whole Group's R&D responsibility to a 37-year-old Silicon Valley insider whose motto is "creative destruction" feels like letting the fox into the henhouse. For sure, both creativity and destruction have been on display since he took the helm. One of the most controversial decisions has been to close the main corporate R&D center near Paris and, in the process, bear the risk of losing (potentially to competitors) tens of key researchers and highly skilled engineers. The move is part of a broader endeavor to recalibrate the profiles of the R&D staff toward new skills such as data analytics, artificial intelligence, and process innovation, but one may wonder whether Airbus, by shaking up its historical and high-profile R&D core, is not throwing the baby out with the bathwater.

Time will tell if the gamble pays off. The biggest part of the wager is not so much the hundreds of millions that are being invested in disruptive ideas and new ventures, but how the overall change of style and direction will impact and shape the organization and its culture. Will borrowing recipes from the Silicon Valley's playbook make Airbus more resilient and successful, or will it damage the DNA of a company whose products stand for reliability and safety, two performance dimensions highly valued by its customers and deeply embedded in the corporate culture?

Changing an organization's culture is no small feat, and a key element of success is how it is accomplished. Going too slowly makes it more difficult to instill new blood without it being diluted in the old one and therefore losing its impact and purpose. Going too fast and too much from the top down, as it seems might be the case for Airbus, bears the risk of employees not "buying it" and resisting the change.

The biggest part of the wager is not so much the hundreds of millions that are being invested in disruptive ideas and new ventures, but how the overall change of style and direction will impact and shape the organization and its culture.

And that is where the biggest challenge lies: Convincing a company with 134,000 employees, a $1 trillion-backlog and a core business based on betting billions on a next-generation aircraft that its future rests on flying cars and cheap mini-satellites, and that $150 million invested in OneWeb or in Airbus Ventures is as relevant to its success as $8 billion of charges taken on the A400M or $15 billion spent on the development of the A350 long-range airliner.

While Airbus' recent innovation initiatives have all the bells and whistles of an organization willing to disrupt itself to remain at the forefront of its industry, it is unclear whether they will have the catalytic effect that Enders hopes will revolutionize the Airbus way of doing business and, by extension, the business of air transport.

As the Buddhist saying goes: *"The finger pointing to the moon is not the moon."*

KEY CONCEPT

CORPORATE VENTURING

Corporate venturing is an increasingly common way for large corporations to invest in start-ups by either setting up their own Corporate Venture Capital (CVC) investment fund or pooling resources with other corporations into a joint CVC fund. There are many possible goals to a CVC initiative:

- **Window on technology** – Accessing and learning about emerging or distant technologies by investing in technology start-ups (technology spin-in) - Forward window on technologies and competitive landscape - Market intelligence / trend detection... – Educating corporate decision-makers on market scenarios and technology options ('eyes and ears').
- **Option building** – Creating growth options by taking stakes in interesting companies or creating new businesses – Reserving the right to play.
- **Leveraging corporate assets** – Unlocking hidden value, exploiting IP by creating new businesses for spin-out.
- **Fostering corporate entrepreneurship** – Encouraging more entrepreneurial spirit in the organization, developing more market- and customer-orientated internal innovator mindset. Create a more entrepreneurial environment that nurtures and rewards speed, teamwork, and prudent risk-taking.
- **Ecosystem development** – Creating an ecosystem that helps broaden and deepen market for the parent organization's products – Investing in "complementor" businesses.
- **Establishing new external links** (with academia, research institutions, VC community...).
- **Encouraging faster and more open innovation** (as part of a broader innovation strategy).

When setting up a CVC fund, key considerations are the field of investment (sectors/types of technology), the geographical footprint and the stage of investment (incubation, seed, early stage, late stage, buyouts...). All three dimensions are related to and conditioned by the investment amount. The smaller the fund, the narrower the field, geographical footprint and stage of investment will have to be.

Chapter 8

NEW SPACE

"Here you leave today, and enter the world of yesterday, tomorrow and fantasy."

Walt Disney, American entrepreneur (1901–1966)

CHAPTER INTRODUCTION

This chapter is a direct follow-on of the previous one, focusing on innovation the space sector and more specifically what is called "New Space", to reflect the development of a new generation of players, coming in with radically new visions, business models and value propositions that are disrupting the traditional "old space" players.

The chapter's first column looks at what is truly disruptive in the space industry, which is not necessarily SpaceX's space launchers – as most observers tend to think, but rather the emergence of miniaturized satellites, which itself has triggered a flurry of new service ventures with original value propositions.

The second column focuses on "master disruptor" Elon Musk's success recipe, which is strikingly close to one used by another great creator and businessman of his time: Walt Disney. It tends to show that, when it comes to great achievements, the field of application matters less than the spirit with which one approaches the challenge.

The third column takes stock of the fact that, while the space sector is in the midst of a major turmoil, with traditional players still fending off new entrants, the writing is on the wall for "old space" players for various reasons, well explained through the lens of Christensen's disruptive innovation theory.

The fourth column assesses the state of the nanosatellite industry and notices that it has not been taking off as anticipated. Following a period of inflated expectations, the sector has now entered a period of uncertainty and transition, which makes it difficult to predict what will happen next.

The last column considers the goal announced by Elon Musk and others to colonize Mars and why a specific type of partnership between the public and the private sectors will be needed to make it happen.

Separating Hype from Reality

March 2014

MUCH HAS BEEN SAID ABOUT the emergence of new players in the space sector and the need for Europe to revamp the way it allocates its resources, starting with the "rule of geographic return." This applies primarily to the launcher segment, where SpaceX's Falcon 9 rocket is perceived as a major threat to Arianespace's commercial supremacy. Suddenly there is a sense of urgency—if only the slightest—in the European establishment.

Should we improve Ariane 5? Should we move to Ariane 6 as soon as possible? How can we get the price of a launch down to $90 million, from more than $130 million today? These are important questions, not least for the thousands of jobs involved. But they miss the point in several respects.

First, SpaceX still has a long way to go to become a fully-fledged competitor in the mainstream commercial launch market. The company is still in the start-up phase and industrializing the business will be a whole new game.

> *SpaceX is a remarkable venture for what it stands for—entrepreneurship, boldness, long-term vision, derring-do—but in the commercial launch market, it is just another player trying to break through.*

In fact, it is not the first time a player has come up with a significantly discounted price. In the early 2000s, Proton and Sea Launch brought launch prices down 40%, taking significant market share from Arianespace. But after several failures, prices went back up to their previous level, and Arianespace was back on top. The impressive track record of Ariane 5 (and of Ariane 4 before it) will not be easy to match, although it is certainly not a reason for Europeans to remain complacent, and it does not change the fact that the Ariane system should be much more cost efficient.

Second, SpaceX is a remarkable venture for what it stands for—entrepreneurship, boldness, long-term vision, derring-do—but in the commercial launch market, it is just another player trying to break through. Even the idea of reusable engines has been thought of before by industry incumbents. To be truly disruptive, an innovation must either lead to a new application outside the mainstream market or to a new business model targeted toward the least-demanding customers. Yet SpaceX has been targeting mainstream customers such as NASA or multinational satcom service providers, with a fairly standard value proposition.

In fact, what is truly disruptive in the space industry is not happening with the launchers themselves but rather with the rapid development of mini satellites. This changes radically the economics of launching objects into space and therefore the kind of technology involved. Hence, many new innovative launch ventures have popped up to accommodate such payloads, among them S3 (Swiss Space Systems), which is designing a shuttle to be launched from the back of an Airbus A300, or GOLS (Generation Orbit Launch Services), which is using a Gulfstream G-III private jet to launch a small missile able to carry a 40-kg (88-lb.) payload, for a target launch price of $1 million.

Other significantly disruptive innovations include the recent development of a standard and modular nanosatellite architecture called CubeSat that allows satellites to be built using almost exclusively commercial, off-the-shelf components, with the possibility to assemble several of them together, for a price below $500,000 apiece.

This innovation itself has triggered a flurry of new service ventures with interesting value propositions, making Google Earth almost look like a dinosaur. California-based Skybox Imaging, for example, raised close to $100 million to launch a new generation of low-cost, sub-meter Earth-observation minisatellites and data analytics services.

These examples provide a reminder that, in the space sector, downstream services are by far where the stakes are the highest: They account for 85% of the overall $100 billion-plus annual space market value, versus 10% for satellites and only 5% for launchers.

In the space sector, downstream services are by far where the stakes are the highest: They account for 85% of the overall $100 billion-plus annual space market value, versus 10% for satellites and only 5% for launchers.

So instead of asking themselves whether they can afford a new launcher or a new $500 million satellite, Europeans should focus their energy and resources on promoting a dynamic service sector. More than anything else, Europe needs dedicated, competitive and innovative service players, which ultimately will offer the only guarantee that there will be a sustainable demand for more satellites and launchers.

Innovation in the space sector is thus not where it seems to be, and competitiveness is certainly not only about the cost of launchers. It is also about making sure new applications are developed, new markets are created, and fully fledged service players are given the support and credit they deserve for making space not only a boxing ring for egos but a truly dynamic and open marketplace.

When Musk Meets Disney

November 2014

SPACEX'S RECENTLY WON $2.6 BILLION contract to supply NASA with a crew transport capsule is a major milestone. SpaceX will become the first private company to launch astronauts into space, 12 years after its creation and six years after being on the brink of collapse. This is a remarkable turnaround and achievement.

What is even more remarkable is that founder Elon Musk seems to be as successful in all his other businesses, be it Tesla or SolarCity, implying he may have found a miracle recipe to break through markets where incumbents—be they defense and space conglomerates, car manufacturers or energy companies—have traditionally crushed potentially disruptive entrants.

So what could this miracle recipe be?

Years ago, I was a consultant to a European start-up that wanted to revive the old Zeppelin's rigid airship design on a large scale for multiple applications, from freight transport to luxury passenger cruises. At first glance, the idea was appealing, to bring back to life a proven technology to offer an environment-friendly, safe, economical, and versatile mode of transportation that could help alleviate road and airport congestion in the busiest parts of Europe. Yet the venture never took off beyond preliminary studies.

In hindsight, four main reasons stand out for such failure: lack of resources, lack of talent, no "system approach" and no "dream" to connect with. Just reverse those causes for failure and you possibly get Elon Musk's recipe for success. The first two, while obvious, are not easy to get. Not everybody can start a company with lots of cash and has the charisma and self-confidence to attract top talent.

The third one relates to the theory of disruptive innovation: as a new entrant, it is extremely hard to disrupt an industry like space or road transportation without creating a whole new eco-system of your own, simply because there are too many vested interests in the existing one. No established space player wants to hear it is possible to reduce the price of launchers by 75%. It is better to keep pushing incremental innovations within the same business model. Musk understood and therefore created his own eco-system, designing his own rocket, vertically integrating his design and production, and acquiring his own test range.

> *The fourth success factor: having a dream to connect to and, as importantly, a process to bridge the gap between the dream and reality.*

Finally, the fourth success factor: having a dream to connect to and, as importantly, a process to bridge the gap between the dream and reality. In that respect, Musk's approach strikes me as similar to one used by another great creator and businessman of his time: Walt Disney. He embodied the ability to create stunning products and a highly successful business by starting with a dream and then applying a systematic and powerful creative process ("storyboarding") to make this dream a reality.

One of the major elements of Disney's unique genius was his ability to explore things from three different perspectives—the dreamer's, the realist's, and the spoiler's—and to build a storyboard by exploiting each perspective iteratively. Musk seems to have been doing just that.

The dreamer asks, "why not?" and sets the dream goal. Why not save humanity by extending its footprint to other planets? Let's send humans to Mars.

The realist asks "how?" How can we send humans to Mars repeatedly and economically? By having reusable space vehicles, which can reduce the cost of spaceflight by two orders of magnitude.

The spoiler asks "yes, but . . . what about?" What about Newton's third law[v] or Tsiolkovsky's equation[vi] that tell us the room for improving a rocket's payload fraction is extremely limited? What about the huge financial resources needed along the way to achieve the necessary technological breakthroughs? How can we keep the cash flowing in? By going where the space money is: satellite launches and transport to the ISS.

This ability to take these different perspectives goes a long way toward explaining Walt Disney's success. As one of his animators noted, *". . . there were actually three different Walts: the dreamer, the realist and the spoiler. You never knew which one was coming into your meeting."* I suspect there

---

[v] **Newton's third law:** For every action, there is an equal and opposite reaction (https://www.grc.nasa.gov/WWW/K-12/rocket/newton3r.html).

[vi] **Tsiolkovsky's equation:** The Tsiolkovsky rocket equation, classical rocket equation, or ideal rocket equation is a mathematical equation that describes the motion of vehicles that follow the basic principle of a rocket: a device that can apply acceleration to itself using thrust by expelling part of its mass with high velocity can thereby move due to the conservation of momentum. The equation relates the delta-v (the maximum change of velocity of the rocket if no other external forces act) to the effective exhaust velocity and the initial and final mass of a rocket, or other reaction engine. http://blogs.esa.int/rocketscience/2012/10/14/a-man-and-an-equation/

are also three Elons, and that whenever he enters a meeting, no one knows which one is coming either.

So while connecting the entertainment world with the space industry may seem farfetched, when it comes to great achievements, the field of application matters less than the spirit with which one approaches the challenge.

When it comes to great achievements, the field of application matters less than the spirit with which one approaches the challenge.

In that respect, Musk could well make Disney's welcoming words his own: "*Here you leave today, and enter the world of yesterday, tomorrow and fantasy.*"

Game Over

October 2015

THE AEROSPACE INDUSTRY IS KNOWN for its long business cycles, which make everything move very slowly. The good thing about it is that you can go off on a yearlong sabbatical and be sure that not much will have changed when you come back. The bad thing is that it makes it hard to assess the quality of strategic investment decisions. Take the Airbus A380 as an example. Even though the program was launched more than 10 years ago, it is still not clear whether it was a bad or good decision, and we may have to wait a decade or more to call it a winner or a loser.

Fortunately, we don't have to wait that long to figure out how the space launch sector is going to play out. Sure enough, the sector is still in the midst of a major shake-up, which makes it difficult to sort out the signal from the noise. Indeed, it seems that not a week goes by without some major announcement or event with the potential to reshuffle the cards. The most recent of those have been the $2 billion bid by Aerojet Rocketdyne to acquire United Launch Alliance (ULA) and the announcement by Jeff Bezos's Blue Origin of its plan to build and launch space rockets from Florida. Is this just background noise, or do these events indicate a deeper shift in the balance of power between the old and new worlds of space?

As for the recent launch failures of Antares and Falcon 9, they may be somewhat of a setback for Orbital Sciences Corp. and SpaceX, respectively, but in reality, they are just part of the "noise" surrounding a period of dramatic change. The "signal" remains the same: A paradigm shift is underway, whereby historical players from the old military-industrial complex are bound to become marginalized and lose out to the new players of the digital economy. Whatever Boeing, Lockheed Martin and Airbus Group do over the next few years, their space launch businesses are unlikely to survive the next decade, for several reasons.

A paradigm shift is underway, whereby historical players from the old military-industrial complex are bound to become marginalized and lose out to the new players of the digital economy.

First, the space launch business is too marginal in the overall portfolio of these groups to receive the required level of investment, attention and resources. They are therefore likely to be overspent and outsmarted by players with more at stake in the business.

Second, they are too isolated in their "space" ivory tower and too disconnected from end-user needs. SpaceX and the like are incubating a whole new ecosystem of like-minded companies in connected areas such as space exploration, space tourism, geo-intelligence and Internet communications. For these players, space is not an end-product; it is an enabler, which means that they are not slaved to any particular system, product or customer. If you compete in your product segment – say space launchers, you don't create growth; you just fight for market shares. Instead, if you compete on what end-users need – say internet connection, you create growth because you compete with other product segments, and that is where success lies.

Finally, their systems, processes, and culture, which have proven so efficient thus far, will lead them to do the opposite of what is required to succeed in the "new world." When your culture tells you to go after proprietary technology, higher margins and economies of scale, you are unlikely to survive in a world of open-source software, low margins and modular architectures. Companies transform themselves successfully not by changing the business models that served them so well in the past but by selling off businesses and starting others. Individual businesses are very hard to change. Think of IBM's transformation in the 1990's. They did not change their core business of making and selling laptop computers. They sold it off and started new businesses, such as cloud computing and software as a service.

The current revolution is particularly worrying for European nations, where a tradition of monopoly and vertical integration by a few large players has stifled innovation and opportunities for smaller players to thrive. This is now backfiring, and suddenly there comes the realization that the industry desperately needs to find new models of competitiveness. But here again, the old world of large vertically integrated, rigidly structured space companies clashes with the new world of networked and agile businesses, happy to experiment with different business models and encourage failure on the road to success.

So whatever the microevents of the next few months or even years—some wins here, some losses there—the current turmoil will sort itself out, a new space order will emerge, and we already know who the ultimate winners will be. Today, it may feel like we are still at a crossroads but, as Woodrow Wilson said, "*you know that where roads come together the separation is small, but where they end the separation is not small.*" Game over.

> **KEY CONCEPT**

BUSINESS MODEL

A business model is meant to encapsulate how a company is able to "monetize" its value proposition. As such it incorporates three main elements: the value proposition itself (what do we sell, to whom and how; basically the "story" the company sells to its customers), the capabilities the company relies upon to deliver the value proposition (i.e., key resources such as people, technology, industrial assets…, as well as key processes such as product development, sourcing, manufacturing, marketing, etc.), and the profit formula (how do we generate revenues, with what cost structure and what margin). A company's business model is thus the "mechanism" that makes these three ingredients (value proposition, capabilities, profit formula) work together in a cohesive and mutually reinforcing way to create a virtuous circle of value creation (for customers) and value capture (for the company).

Here below is an illustration of a business model for a defense technology company, in which a strong incremental innovation capability generates a regular stream of new, 'high performance' products thus maintaining customer loyalty and entry barriers, which confers the company significant pricing power. Such pricing power, combined with a well-thought price skimming strategy, allows the company to consistently generate high profits, which can then be reinvested to feed the innovation engine.

In the early stages of a company's existence, resources tend to be fluid and processes under- developed, making it relatively easy for the company to tweak its business model as it goes (based on market feedback and competitive dynamics), by changing either its value proposition, capabilities, or profit formula. That is why start-ups often experiment with different business models before settling on a specific one.

But once a company matures and finds a business model that works for a given value proposition, it tends to allocate its resources and develop processes in a way that supports and nurtures that specific business model.

The key then is to make sure the value proposition remains relevant to the target customers. If for any reason, a mismatch occurs between what customers value and the firm's value proposition (for example price vs. reliability), then it can derail the profit formula and make the historical business model obsolete or counterproductive. That is partly what is happening in the space launcher sector. As commercially driven innovations increase the supply of launch solutions and bring launch prices down with an acceptable level of reliability for many customers, the business model of the incumbent players (essentially a "cost-plus" model in which the only way to sustain profits is to offer always more reliability and complexity) is being severely challenged.

In such context, the only way for incumbent players to stay relevant is to change their value proposition (and business model) so that it better matches the market needs; but because their capabilities are tied to a given business model, they often find themselves incapable or unwilling of doing so, at least until they do not have any other choice, but by then it is likely to be too late as customers will have already moved on.

BUSINESS MODEL ILLUSTRATION
(Defense technology company)

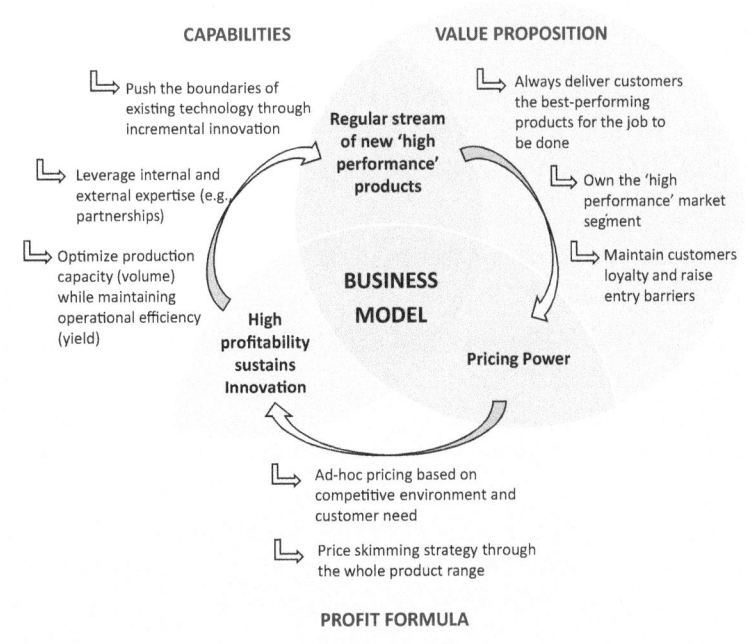

CAPABILITIES
- Push the boundaries of existing technology through incremental innovation
- Leverage internal and external expertise (e.g. partnerships)
- Optimize production capacity (volume) while maintaining operational efficiency (yield)

VALUE PROPOSITION
- Always deliver customers the best-performing products for the job to be done
- Own the 'high performance' market segment
- Maintain customers loyalty and raise entry barriers

Regular stream of new 'high performance' products

BUSINESS MODEL

High profitability sustains Innovation

Pricing Power
- Ad-hoc pricing based on competitive environment and customer need
- Price skimming strategy through the whole product range

PROFIT FORMULA

Source: Author

Nanosatellites: Time to Deliver

June 2019

THE "NEW SPACE" REVOLUTION has been in the making for more than a decade but 2014 was a particularly pivotal year for new satellite companies: Google acquired start-up Skybox Imaging for $500 million and Planet Labs (now Planet) deployed its first two Dove 3U CubeSats from the International Space Station. Both Skybox and Planet were founded in the Silicon Valley a few years earlier with a vision to revolutionize access to information generated from timely, very high-resolution satellite image data with cutting edge data extraction and analysis.

From then on, it looked like it would be a matter of years before low Earth orbit would be filled with thousands of micro- or nanosatellites (most of them in the CubeSat format) and tens of commercial constellations offering services, from real-time images of the Earth to the networking of millions of smart sensors around the globe. Venture Capitalists and other investors grabbed the opportunity and billions of investment money followed.

Five years on, it is time for a reality check. A recent study by Paragon European Partners offers a sobering view of the state of the nanosatellite industry. In a nutshell: the hype is over.

Today, only two commercial nanosatellite constellations are operational: Planet and Spire Global. Together they account for 90% of all commercial nanosatellites launched (excluding academic, scientific, and institutional satellites), the total of which was a mere 500 or so at the end of 2018. While these two constellations are up and running, they are still falling short in terms of service performance to the point where Planet had to make a "pact with the devil" by partnering with Airbus, the very player it was meant to disrupt. As for Spire, it is still trying to figure out how to properly monetize its growing fleet of CubeSats.

Across the Atlantic, European nanosatellite start-ups have also been proliferating, hoping to join the gold rush. Denmark's GOMspace, recognized as a world leader in CubeSat technology, was listed on the Stockholm Stock Exchange in 2016. Its valuation reached more than $400 million at its peak in 2017. Eighteen months later the stock has lost more than 80% of its value and the company is bleeding cash and generating just a little over $10 million in annual revenue. Meanwhile, Australia-listed Sky and Space Global, a relatively high-profile Internet of Things (IoT) constellation company with three satellites in orbit (and a GOMspace customer), has experienced a similar collapse. Its stock value has fallen by 90% (almost $400 million) since its 2017 peak (*see chart below*).

THE BURST OF THE 2016-2017 NANOSATELLITE SPECULATIVE BUBBLE

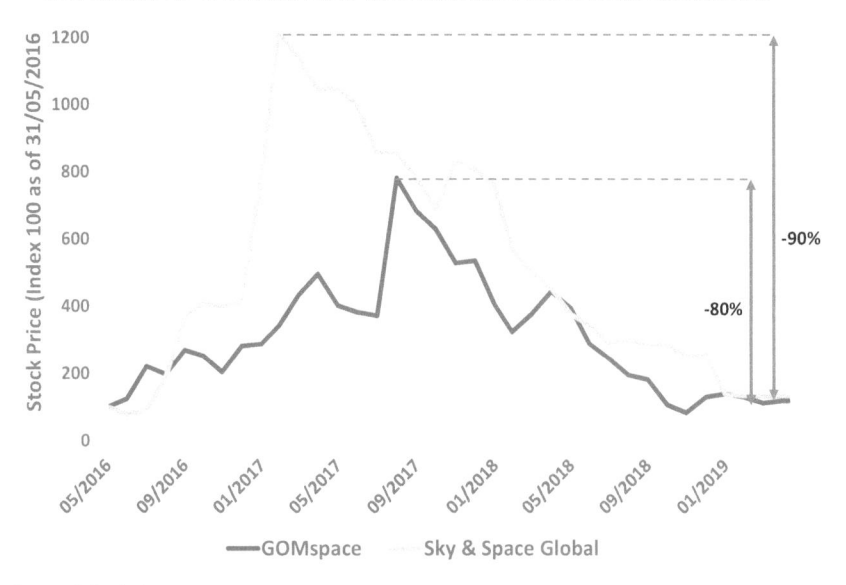

Source: Yahoo! Finance

Clearly something has gone wrong with the nanosatellite revolution. Or has it? In fact, all indicators point to the industry being in a phase that technologists call the "trough of disillusionment": Following a period of inflated expectations, the sector has now entered a period of uncertainty and transition from the visionary phase to the pragmatic phase of innovation and market adoption (as per Geoffrey Moore's model). Paragon estimates the current industry maturity level at around 15%, which means that the actual level of industry revenues is 15% of what it should generate based on the theoretical market size (which is itself based on all the announced constellation projects).

The reason for such low maturity level is that commercial constellations have by and large not yet materialized. Out of 50 or so announced projects (excluding Planet and Spire) that should lead to a total number of 3,000 nanosatellites in orbit over the next few years, none has launched more than a few satellites, let alone started to mass-produce them, and only a handful have raised enough money to be confident of doing so.

> *Following a period of inflated expectations, the sector has now entered a period of uncertainty and transition from the visionary phase to the pragmatic phase of innovation and market adoption.*

Yet, the market can only take off if there is a critical mass of satellites to be produced annually, thus generating economies of scale and driving prices down. As an illustration, the target price of a 3U CubeSat produced in small volume (say 100 per year) is about $500,000. But today, the actual market price is closer to $2 million, if not more.

> *The market can only take off if there is a critical mass of satellites to be produced annually, thus generating economies of scale and driving prices down.*

So, what's next for the nanosatellite industry? Are we going to reach the next stage of the technology maturity curve, so-called the Slope of Enlightment, and finally see the disruptive business model of the "New Space" start-ups vindicated? Or will the speculative bubble of 2016–17 be remembered as a false start for the mass-commercialization of space with miniaturized satellites?

Certainly, the next couple of years are going to be critical for all the commercial constellation projects and the manufacturers behind them. The hype is over. Time to get down to business and deliver on the promises made to investors and customers.

KEY CONCEPT

TECHNOLOGY CURVE / INNOVATION CYCLE

Two models can help us understand what is happening with the nanosatellite industry. The first one is Gartner's Technology Maturity Curve. It breaks down the technology's life cycle into five typical phases:

1. **Innovation Trigger:** A potential technology breakthrough kicks things off. Early proof-of-concept stories and media interest trigger significant publicity. Often no usable products exist, and commercial viability is unproven.
2. **Peak of Inflated Expectations:** Early publicity produces some success stories — often accompanied by scores of failures. Some companies take action; many do not.
3. **Trough of Disillusionment:** Interest wanes as experiments and implementations fail to deliver. Producers of the technology shake out or fail. Investments continue only if the surviving providers improve their products to the satisfaction of early adopters. This is where the nanosatellite industry seems to be right now.
4. **Slope of Enlightenment:** More instances of how the technology can benefit the enterprise start to crystallize and become more

widely understood. Second- and third-generation products appear from technology providers. More enterprises fund pilots. Conservative companies remain cautious.
5. **Plateau of Productivity:** Mainstream adoption starts to take off. Criteria for assessing provider viability are more clearly defined. The technology's broad market applicability and relevance are clearly paying off.

The second model is **Geoffrey Moore's interpretation of the standard Innovation Adoption Cycle.** The standard model breaks down the cycle into five categories of customers whose numbers follow a bell curve: Innovators, Early Adopters, Early Majority, Late Majority, and finally Laggards.

Moore demonstrates that in fact, there are cracks in the curve, between each phase of the cycle, representing a disassociation between any two groups; that is, "the difficulty any group will have in accepting a new product if it is presented the same way as it was to the group to its immediate left."

The largest crack, so large it can be considered a chasm, is between the Early Adopters and the Early Majority. Many (most?) high tech ventures fail trying to make it across this chasm. It looks like the nanosatellite industry is in the middle of it.

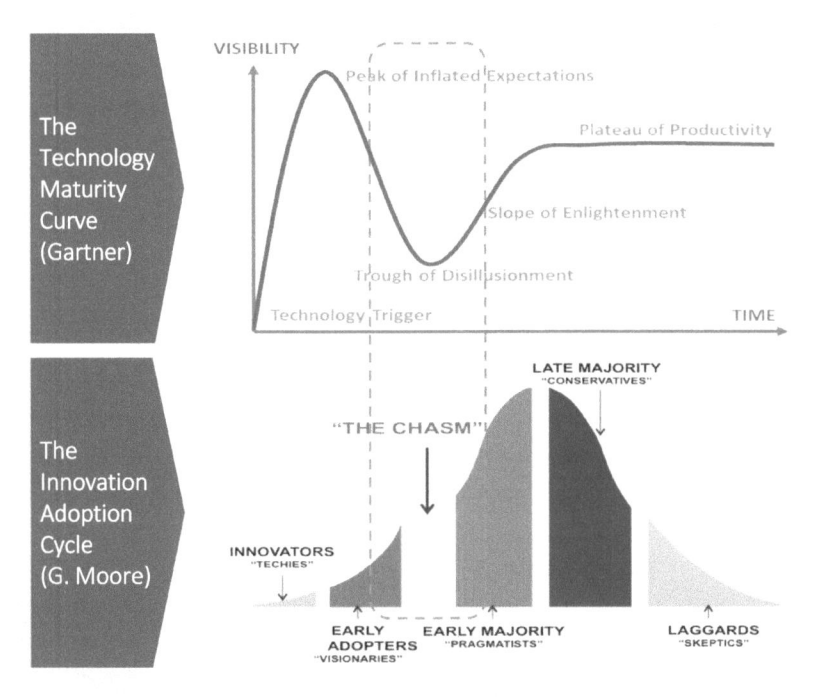

Source: Gartner's Technology Maturity Curve, Geoffrey Moore's Innovation Adoption Cycle

Colonizing Mars

October 2016

Within a month, the aspiration to send humans to Mars seems to have reached a new level of media exposure. First Jeff Bezos's Blue Origin disclosed its plan to build the New Glenn, a rocket with the potential to send humans into space. Then SpaceX CEO Elon Musk presented his vision of how we could shuttle to and from Mars within a couple of decades. And two weeks later, President Obama wrote an op-ed calling for America to set its sights on sending humans to Mars by the 2030s with the ambition of remaining there for an extended time.

While coming from different angles, both Musk and Obama emphasized the need for a public-private partnership to achieve these ambitious goals.

Musk's main objective is to make the trip to Mars affordable for as many people as possible. His hypothesis is that if one can bring the cost down to the median cost of a house in the US—$200,000—then there will be a critical mass of people who can afford and are willing to go. In order to reach that affordability threshold, he believes government money will be needed along the way, hence the need for a public-private partnership.

Obama's main purpose is to stimulate innovation, inspire a new generation of scientists and engineers and ultimately revive his country's pioneering spirit that led it to win the first space race to the Moon. He also believes a partnership between government and the private sector will be beneficial, if only to harness and support America's entrepreneurial ingenuity and derring-do. Yet, in spite of these nice intentions, it is worth asking what a public-private partnership would mean in such a context.

Cooperation between the private and public sectors in the aerospace industry is usually about one financially supporting the other—e.g., private institutions helping fund government projects through such mechanisms as a Project Finance Initiative or public money funding private projects through R&D contracts. In the case of establishing a long-term presence on Mars, though, funding is unlikely to be the main obstacle. If anything, judging by the amount of cash and level of indebtedness, the private sector is today wealthier than the public sector, and the financial resources of leading technology companies like Google, Amazon or Microsoft dwarf those of NASA. So it could well be that the private sector alone could finance the effort. However, this does not mean that cooperation with the public sector is not needed. In fact, it is essential, but not mainly for financial reasons.

As for any new venture, supply and demand need to meet somewhere to achieve success. The technology zealots, internet billionaires, corporate giants and NASA will take care of the supply of rockets and space infrastructure, and the race is already on to be the first to transport a human to Mars. But what about demand? I am not talking about demand from people willing to pay to go to Mars; this will likely be marginal compared to the total costs of going and settling there. What one must consider instead is the "demand" for colonizing Mars in order to make humans a multi-planet species and thus mitigate any risk of extinction on Earth. For the project to be viable, there needs to be a critical mass of people buying into the vision and the idea that it is worth it.

For that purpose, going to Mars has to be sold not as a luxury cruise for space tourists but as a mission of public interest and—possibly—of universal humanitarian value, and this will require a significant amount of public resources.

> *Going to Mars has to be sold not as a luxury cruise for space tourists but as a mission of public interest and—possibly—of universal humanitarian value.*

This is where Musk's approach is flawed. If he truly believes that what is at stake is the long-term survival of humanity, then he should not expect to charge people $200,000 for the trip. Instead—and that is why a true private-public partnership is needed—it should be advertised as a public-service job to help establish a human colony on Mars for which people will be paid. The application process would then be more akin to enrolling in the Army, or more fittingly, in the Peace Corps.

In fact, it would be quite appropriate, and visionary, for the US government to start thinking along those lines and, for example, launching a kind of "Space Corps" organization whose purpose could be directly inspired by the Peace Corps Act of 1961.[vii] It could then read something

---

[vii] **The Peace Corps Act of 1961:** The Peace Corps is a volunteer program run by the United States government. Its official mission is to provide social and economic development abroad through technical assistance, while promoting mutual understanding between Americans and populations served. On March 1, 1961, President John F. Kennedy signed Executive Order No. 10924, establishing the Peace Corps as an agency in the US Department of State, and later that same year Congress adopted the Peace Corps Act (P.L. 87-293): *"to promote world peace and friendship through a Peace Corps, which shall make available to interested countries and areas men and women of the United States qualified for service abroad and willing to serve, under conditions of hardship if necessary, to help the peoples of such countries and areas in meeting their needs for trained manpower, particularly in meeting the basic needs of those living in the poorest areas of such countries, and to help promote a better understanding of the American people on the part of the peoples served and a better understanding of other peoples on the part of the American people"*. Since its inception, more than 235,000 Americans have joined the Peace Corps and served in 141 countries.

like this: *"[The Space Corps] shall make available to Mars and other planets men and women of the United States qualified for service in space and willing to serve, under conditions of hardship if necessary, to help establish a human multiplanetary civilization."* Join today! Mars needs you!

Chapter 9

THE BIG(GER) PICTURE

"It was that quality that led me into aviation in the first place, it was a love of the air and sky and flying, the lure of adventure, the appreciation of beauty."

Charles Lindbergh, American aviator (1902–1974)

CHAPTER INTRODUCTION

This final chapter is meant to step back from the purely business aspects of aerospace and addresses what I regard as more fundamental questions about the future of the industry, worth reflecting upon.

First, I notice that air travel is probably one of the very few sectors—if not the only one—in which service standards seem to have been frozen in time. I therefore question whether hypersonic technology, for example, is the kind of innovation the industry needs today. Does "faster" necessarily mean "better"?

In the second column, I raise the BIG issue of air travel sustainability and argue that it is an illusion to believe that supply-driven measures will be enough to achieve the industry's ambitious zero-emission goals. In that respect, serious progress will not be achieved unless we tame the aviation's "sacred cow": demand growth.

The third column is looking at the new urban air mobility concepts and how they could take the aviation industry in a completely new direction, reversing the race for "bigger, faster, farther" that has been the dominant paradigm for almost a century. Instead, "smaller, slower, closer" might well become the new paradigm of air travel.

The fourth column raises the question of the purpose of space exploration and the possibility that space exploration is not only about technological breakthroughs or scientific knowledge, but also about going deeper into our own psyche and our understanding of our human condition.

The final column, written before the outbreak of the COVID-19 pandemic, connects Boeing's 737Max crisis and the aerospace community's sustainability challenge and sees both as a sign that commercial aviation may be entering a new age, defined not anymore by technological prowess, corporate wealth accumulation or relentless growth, as has been the case for the last 30 years, but by restraint, humility, and resilience.

The Lost Art of Air Travel

August 2015

THE RECENT AWARD TO AIRBUS of a patent for a hypersonic jet has revived discussion of the feasibility of and interest in a Mach 5+ aircraft that could take passengers from Paris to Tokyo in 3 hours. Several articles about it have appeared in this magazine, highlighting the underlying technical and operational challenges. Of particular concern are the economics and environmental impact of flying such an aircraft.

Regardless of these challenges, the aerospace community seems to agree that increasing the speed of flight so drastically is the kind of achievement aerospace engineering is all about and, therefore, it is worth investing in. After all, reducing to 3 hours what currently takes more than 11 hours would be major progress.

But would it be progress, really? Perhaps it would actually be a misallocation of resources. Instead of investing in faster aircraft, what about investing in something that would truly make a difference for most air travelers, such as service quality and customer experience?

I have been a frequent flier for 30 years, across all classes of air travel and types of airlines—including a stint as a flight attendant during my university years. It seems to me that, while the quality of service in most industries has improved significantly in the last few decades (thanks to technology and increased focus on customer needs), air travel is probably one of the very few sectors—if not the only one—in which service standards seem to have been frozen in time.

Air travel is probably one of the very few sectors in which service standards seem to have been frozen in time.

To be clear, I am not talking about reliability or efficiency (which had to improve to cope with the sheer increase in the number of aircraft and passengers) but about customer experience. I am shocked by how onboard service has remained largely unchanged since my early flying years. Airlines are using the same contrived service routine, onboard service equipment, safety announcements and cabin configurations. I am disturbed by the unreliability of supposedly high-tech airport security systems that will every so often ring without reason, thus subjecting me to the probing of a well-meaning but sometimes overzealous agent.

I dread the time I must spend walking through the maze of shops strategically laid out between airport security and departure gates, harassing my senses with blends of dubious fragrances. Overall, I am dismayed by how air travel has become such a mechanical, near-robotic experience in which the goal is to "process" as many passengers as possible in the shortest time and cram them into the smallest space while maximizing commercial revenues throughout their "captive" presence.

Under the banner of "safety and security first," each step of the journey is scripted down to every word and gesture. All flight attendants say the same words; every passenger is meant to perform the same gestures when requested. As American psychologist Barry Schwartz notes, scripts like these are insurance policies against disaster and, by and large, they do prevent disasters. But what they also ensure, when applied too mechanically, is mediocrity in human interactions and in the overall travel experience.

So I do question whether hypersonic technology is the kind of innovation the industry needs today. Surely, wealthy individuals already flying private jets or in first class on commercial airliners may be interested in speeding up their journey. But for most of us, there is much more at stake than saving time—such as flying in more energy-efficient and environmentally friendly aircraft or escaping the commodity trap that air travel has fallen into and making these 9 or 12 hr of flight an opportunity for quality time instead. For some people that may mean doing some work; for others it will mean relaxing or entertaining themselves. For all of us, it should mean reflecting on the fact that distances—be they geographical or cultural—are things to be relished rather than minimized.

If you want your organizations to innovate in a meaningful way, if you want to really contribute to the progress of your industry, what about setting yourself a mission of introducing some humanity and art back into the business of air travel?

In a nutshell, my rallying cry to aviation and aerospace executives is this: If you want your organizations to innovate in a meaningful way, if you want to really contribute to the progress of your industry, what about setting yourself a mission of introducing some humanity and art back into the business of air travel? What about investing your R&D money in creating environments more conducive to enjoyable and truly memorable experiences?

Indeed, for a service business, the ultimate sign of success is not that your customers wish to "get out of here" as quickly as possible; it is that they don't mind remaining in your company a little longer because they are enjoying themselves so much.

Dirty Secret

August 2016

WHEN IT COMES TO SUSTAINABILITY of air travel, the aviation community is lying to itself. There is an elephant in the room, and nobody wants to see it. Without more significant efforts—some would call them sacrifices—air travel will continue to be a huge contributor to this planet's pollution and global warming, with dire consequences for future generations, whether there are air travelers or not.

For the first time last month, the US Environmental Protection Agency (EPA) has officially acknowledged its duty to promulgate standards applicable to greenhouse gas (GHG) emissions from commercial aircraft. Indeed, aircraft remain the single largest GHG-emitting transportation source not yet subject to GHG standards in the US The uptake of sustainable alternative fuels, which are meant to be a large part of the solution by 2050, has been extremely slow, with only two airports in the world—Oslo and Los Angeles—offering biofuels to airlines.

As for the International Civil Aviation Organization's (ICAO) progress on setting a CO2 emissions standard, it is as slow as it can get. Having set some "aspirational" goals in 2010 for global net carbon emissions, it is only now getting to the point of recommending that new aircraft models entering service after 2020 and existing aircraft models coming off the production line after 2023 meet the new technical standard—which has yet to be introduced.

The bottom line is that with all the talk about the aviation community being committed to action on climate change (see, for example, ATAG's position paper[viii] signed by aviation industry leaders in 2012) and about how various supply-driven measures will improve air travel's sustainability, everybody knows that serious progress will not be achieved unless we address the other part of the equation, aviation's "sacred cow"—demand.

> While current measures to improve air travel's sustainability are all supply-driven, serious progress will not be achieved unless we address the other part of the equation, aviation's "sacred cow"—demand.

[viii] ATAG Position Papers: https://www.atag.org/our-publications/latest-publications.html

Air travel has essentially doubled in the past 15 years (and so have related CO_2 emissions) and is expected to double again over the next 15 years. There is just no way supply-based measures such as technology and infrastructure improvements will come close to offsetting such rapid growth and its impact on the environment. As for the so-called "market-based measures," such as the European Union's Emissions Trading System or other carbon-offsetting schemes, they are just gap-filling measures that make people feel good without really changing the fundamental dynamics of the industry.

These dynamics are essentially about convincing as many people as possible to fly as often as possible. This is achieved in two ways: First, by making air travel affordable for mass consumption and second, by enticing business travelers with all sorts of bells and whistles.

The low-fare airline model has certainly made air travel more accessible to many, but in that process, it has contributed to its excessive commoditization. And while it may be "low fare," it certainly is not "low-cost" as far as the environment is concerned. In that respect, air travel is very much akin to fast food: It may seem to be a bargain, but it passes the true cost on to the public health and purse and pushes it into the future.

As for major airlines, they are just milking the top of the market by catering to the desires of an international business community that is self-important enough to believe it deserves nothing less than increasingly dedicated and expensive services such as all-business-class flights, upscale cabins, exclusive lounges and loyalty rewards. Yet frequent flier programs themselves create the wrong incentives, as they encourage business customers—most of which do not pay for their tickets themselves—to fly more and spend more on flights than may be necessary.

The low-fare airline model has certainly made air travel more accessible to many, but in that process, it has contributed to its excessive commoditization.

If we in the aviation community are really serious about addressing climate change, we first need to accept that the current and forecasted demand for air travel is unsustainable and therefore must be dampened. We as air travelers must become wiser customers: Do we actually need to fly to all these business conferences? Is it worth paying 10 times the price of an economy ticket for better wine, free lounge food and a more comfortable seat? Do we really believe a €20 ($22) trip from London to Copenhagen covers its true cost and true value?

Travelling to different countries can be one of the most enriching experiences in life. Let's not trivialize it to the point where we do not think twice before buying a cheap air ticket. Let's be honest with ourselves and accept the basic but uncomfortable truth that to make air travel sustainable, we first need to change our individual behaviors and become more discerning consumers. In the must-win fight for sustainability, there cannot be gain without pain.

Smaller, Slower, Closer

January 2017

THE BEGINNING OF A NEW YEAR is always a good occasion to reflect on major trends that have been taking shape in the past 12 months. One development in commercial aviation is the emergence of new air transportation concepts taking us one step closer to a "Blade Runner" type of future, where flying cars become part of a three-dimensional transportation system.

Uber's Elevate concept paper[ix], published in October, gives a glimpse of what is on the drawing board: fleets of autonomous vertical-takeoff-and-landing (VTOL) vehicles flying around megacities and cutting our daily commuting time by nearly 90% for a price that could ultimately be the same as a taxi ride.

Regardless of how and when this vision becomes reality, there is no question that such futuristic transportation concepts have the potential to take the aviation industry in a completely new direction, with some major consequences for all its stakeholders.

The race for bigger, faster, and farther may finally be reversed.

The most obvious one is that the race for bigger, faster and farther may finally be reversed. From the time when aviation pioneers like Charles Lindbergh were flying single-seat, single-engine, purpose-built monoplanes to the current generation of mass-produced jets that can carry several hundred passengers, the history of commercial aviation has been primarily about creating bigger and more powerful flying machines. In that race, Boeing and Airbus have come out on top, monopolizing the market for large commercial aircraft, and setting the innovation agenda for the rest of the industry.

Today, however, the challenge for new air transportation systems is of a quite different nature. In the context of on-demand urban air transportation, it is not about making "bigger" better but about making "smaller" better—both technologically and economically.

Technologically, the constraints for flying in an urban environment are such that only vehicles of a certain size will be able and allowed to do it. And

---

[ix] **Uber Elevate Concept Paper:** https://www.uber.com/elevate.pdf

they will need to be quiet, fast, clean, efficient, and safe, all of which is more likely to be achieved with technologies such as electric propulsion, vertical lift and autonomy that are better adapted to small-scale vehicles. Even solar propulsion could be an option for some small vehicles, as Bertrand Piccard's Solar Impulse first round-the-world solar flight brilliantly demonstrated last year.

As far as autonomy is concerned, thanks to swarm intelligence and advanced precision positioning technologies, it may actually turn out to be easier and safer to operate and monitor fleets of small, self-driving air vehicles than large transcontinental airliners flying over oceans.

Economically, the only way to make such new means of transportation viable—even in a ridesharing or taxi format—is to bring the acquisition price of the vehicle close enough to that of a typical car. This will require the vehicles to be small enough, if only to fully leverage the potential of 3-D-printing technology, and to make thousands, if not tens of thousands, of them every year to generate sufficient economies of scale.

Overall, this points to a future of aviation in which most innovations will be driven by new applications and new performance dimensions (such as noise, emissions, and agility), where the economics of manufacturing VTOL vehicles will become more akin to automobiles than aircraft, and flying will become much more user-centered and part of a fully integrated and largely individualized mobility service.

The aviation ecosystem as we know it is therefore likely to change dramatically in the next few decades. As people start rethinking transportation in three dimensions and flying becomes an integral part of individual mobility scenarios, traditional air travel—on large airplanes, to and from distant airports, on fixed schedules and limited routes—will increasingly look old-fashioned and out of sync with the modern urban lifestyle. Airlines may rebrand themselves as luxury cruise operators, Boeing and Airbus may keep battling it out, but they will lose their status of industry trailblazers. The aviation world as a whole will become much more diverse and its boundaries much more permeable.

> *As people start rethinking transportation in three dimensions and flying becomes an integral part of individual mobility scenarios, traditional air travel will increasingly look old-fashioned and out of sync with the modern urban lifestyle.*

Ultimately, thanks to technology and human creativity, aviation in the 21st century may even bring us back to the original vision that drove many of its pioneers but was widely lost in the quest for largeness and productivity: make air travel suitable for small-scale applications, cheap enough so that it is accessible to virtually everyone and compatible with the human aspiration for freedom, adventure, and beauty. As Lindbergh said: *"It was that quality that led me into aviation in the first place, it was a love of the air and sky and flying, the lure of adventure, the appreciation of beauty."*

WHAT IS THE PURPOSE OF SPACE EXPLORATION?

SEPTEMBER 2018

NEVER BEFORE HAS SPACE EXPLORATION attracted so much attention and so many resources across the whole spectrum of society, from astrophysicists to movie directors, from governments to private start-ups and from billionaire investors willing to bet on asteroid mining to college classrooms where students are building their own CubeSats. Almost five decades after the first Moon landing, a whole new ecosystem is taking shape to enable not only a return to the Moon but also the establishment of a cislunar infrastructure and beyond that a human settlement on Mars.

One of the main catalysts for such revival has undoubtedly been Elon Musk's creation of SpaceX in 2002, with its stated mission to make humans a multi-planet species by sending them to Mars. Sixteen years later, so much has happened that what seemed like an outrageous, ego-driven proposition has become a tangible and respected enterprise, endorsed by the international space establishment as well as the investor community, for which the risk/return ratio is now much more palatable. The biggest endorsement of all has come from NASA, which has set itself the strategic goal of "returning American astronauts to cislunar space and the Moon to build the foundation we need to send Americans to Mars and beyond." Suddenly, the Moon is not an end anymore but just a stepping-stone toward farther targets, a kind of base camp for deeper space exploration.

Almost five decades after the first Moon landing, a whole new ecosystem is taking shape to enable not only a return to the Moon but also the establishment of a cislunar infrastructure and beyond that a human settlement on Mars.

And so, with such boundless perspectives and exciting developments, for the first time in many decades "space" is stealing the limelight away from its "aero" counterpart in the overall aerospace industry, which I suppose is welcome news for those tired of watching the Boeing-Airbus ping-pong match. More seriously, this new focus on space exploration is also the opportunity for the aerospace community to reconnect with a sense of higher purpose that has largely been lost over the years, as the industry has become mainly utilitarian and focused on operational efficiency and market competitiveness, be it in commercial aviation or satellite communications.

But what is the purpose we are talking about? Is it what Musk calls our "duty to maintain the light of consciousness"? Is it to search for extra-terrestrial intelligence and thus validate astronomer Frank Drake's equation[x] as to the probability of other observable civilizations in our galaxy? Is it to find out whether inventor Nikola Tesla and quantum physicist David Bohm were right in stating that space is not empty but filled with a kind of force field or "cosmic plenum" that holds everything together? Or is it simply to find something about ourselves that is only accessible from far away?

Several former astronauts have already explained how looking at Earth from space opened up new perspectives on life and the planet. As Apollo 14 astronaut Edgar Mitchell expressed it coming back from the Moon: *"You develop an instant global consciousness, a people orientation, an intense dissatisfaction with the state of the world and a compulsion to do something about it."* Similarly, we can expect that going deeper and further in our understanding and use of space will raise new questions about the Universe and our position in it, and it might challenge some fundamental assumptions underlying our civilization.

Mitchell also said: *"Looking beyond the Earth itself to the magnificence of the larger scene, there was a startling recognition that the nature of the Universe was not as I had been taught. . . There was an upwelling of fresh insight coupled with a feeling of ubiquitous harmony—a sense of interconnectedness with the celestial bodies surrounding our spacecraft."*

Comments such as these from a NASA astronaut may sound out of place in a primarily scientific and materialistic context, but they resonate with what quantum science already teaches us about the connection between mind, matter, and the cosmos, and they hint at the possibility of a significant space-induced leap in human consciousness as we collectively become a spacefaring civilization.

Space exploration is not only about technological breakthroughs or scientific knowledge. It also will almost certainly take us deeper into our own psyche and our understanding of human consciousness.

Thus, space exploration is not only about technological breakthroughs or scientific knowledge. It also will almost certainly take us deeper into our own psyche and our understanding of human consciousness. It is not only a quest

---

[x] **Drake Equation:** In 1961, astrophysicist Frank Drake developed an equation to estimate the number of advanced civilizations likely to exist in the Milky Way galaxy. https://exoplanets.nasa.gov/news/1350/are-we-alone-in-the-universe-revisiting-the-drake-equation/

for what is "out there;" it is also a path to discover what is "in here." And it is certainly not about escaping our human condition but rather about becoming more human and more in tune with our inner being. *"We went to the Moon as technicians; we returned as humanitarians,"* Mitchell said. This prospect itself makes space exploration a worthwhile investment and the decades ahead something to look forward to.

The Age of Resilience

October 2019

MY JOB, AS A CONSULTANT, is basically to connect the "dots" to find out patterns and trends. This week, two seemingly unrelated "dots" attracted my attention. The first one was a mass email from Boeing with the header "New Steps to Strengthen Airplane Safety," which listed 10 actions taken by the company to "strengthen the culture of safety throughout Boeing and the broader aerospace industry" in the wake of the 737 MAX crisis. The other was a report[xi] published by Imperial College London titled "Behaviour Change, Public Engagement and Net Zero," which essentially states that the only way to reach the UK's net-zero emissions 2050 target is to constrain demand for air travel (on top of all other supply-based measures).

On the face of it, they had nothing to do with each other: safety and sustainability, one aerospace company and the whole aviation industry, a short-term problem and a long-term one. Yet there was a clear connection between the two: They both dealt with the two sacred cows of commercial aviation—safety and (demand) growth—and they both pointed to a similar moment of truth.

Boeing's email is fascinating in several ways. It gives a glimpse of the trauma the company has been experiencing since it was caught in the act of failing in its core mission of designing and delivering safe airplanes. It also shows how desperate Boeing is to make up for it by launching all sorts of committees, programs and systems meant to improve the way its products are made. Unfortunately, the email also comes across as clear evidence that Boeing has not identified the root cause of the problem, or if it has, the company is not sure about how to tackle it.

Commercial aviation may be entering a new age, not defined by technological prowess, corporate wealth accumulation or relentless growth, as has been the case for the last 30 years, but rather by restraint, humility and resilience.

A clue to the root cause can be found in the use of the word "culture" (of safety) in the introduction of the email. Culture is fundamental because it drives behavior. Culture is to companies what character is to individuals: It shapes and gives

---

[xi] **Report for the Committee on Climate Change** (Imperial College London, October 2019): https://www.theccc.org.uk/wp-content/uploads/2019/10/Behaviour-change-public-engagement-and-Net-Zero-Imperial-College-London.pdf

direction to their actions, and something obviously went wrong for Boeing in that respect. Has the company become too big to manage? Has the focus on profitability and market share pushed people to cut corners? Is it just the complacency and hubris that often come with success?

Whatever it is, putting in place committees, systems and processes will not solve the whole problem, because at the end of the day, it comes down to individual behavior and people's ability (and willingness) to change their view of the world.

And this is also what Imperial College's report is about. As the UK has set itself the ambitious goal of net-zero emissions by 2050, and as aviation is now clearly identified as the likely largest emitting sector by then, it is now widely accepted—as I wrote in this column three years ago—that supply-driven measures (such as aircraft design, fuel efficiency, alternative fuels, etc.) will not be enough. Demand must be brought under control.

There are various ways to do this, such as taxing airline fuel or airline tickets (like France will start doing next year) or, as the report recommends, implementing an escalating air-miles levy to discourage excessive flying by the 15% of the population estimated to be responsible for 70% of flights, or even banning frequent-flier reward schemes. Whatever the means, the key message is that individuals must be nudged to change their consumption behavior if any impactful air travel sustainability outcome is to be achieved.

And so, as Boeing and the broader aerospace community struggle to come to terms with quite possibly their biggest challenge yet, one must recognize that commercial aviation may be entering a new age. This new age will be defined not so much by technological prowess, corporate wealth accumulation or relentless growth, as has been the case for the last 30 years, but rather by restraint, humility and resilience: **restraint** in not overpromising on technical performance to the detriment of safety; **humility** in accepting that perhaps we overdid it by making flying so cheap that the true cost of it has been transferred to nature; and **resilience** because whenever we are going through a trauma, be it an organizational one like Boeing's or a societal one like climate change, we must find the resources and courage to question the paradigm we have been accustomed to and

> *Whenever we are going through a trauma, be it an organizational one like Boeing's or a societal one like climate change, we must find the resources and courage to question the paradigm we have been accustomed to and ultimately change our individual behavior to spring back to a collectively sustainable state.*

ultimately change our individual behavior to spring back to a collectively sustainable state.

Only by embracing such a mindset will we be able to stop sacrificing the essential to urgency and instead start addressing what French philosopher Edgar Morin calls "the urgency of the essential."

"As we keep sacrificing the essential to urgency, we forget the urgency of the essential"

Edgar Morin

Conclusion

While the COVID-19 crisis has been dealing a huge blow to the industry and has the potential to significantly reshuffle the cards, the stories covered in this book and the lessons and insights drawn from them are, I believe, pointing a way forward. As the industry's foundations are shaken and its weaknesses laid bare, it is critical for all stakeholders to step back and think about the structural factors that will allow the industry and its constituents to be successful again, but this time in a more sustainable way. It entails avoiding the pitfalls of the past and setting a new agenda for a more resilient future.

This concluding chapter lays out such agenda, which – as it turns out – almost reads like a summary of many of the themes covered throughout the book. This is because what will be needed first and foremost from corporate leaders is to go back to the strategy drawing board, looking at the forces at play and sorting out the signals from the noise, which is essentially what this book is about.

All stakeholders have a role to play and should contribute to make the overall ecosystem more resilient: Industrial players (large and small), governments (including agencies and regulatory bodies), investors (private equity and capital markets) and customers (institutional and commercial).

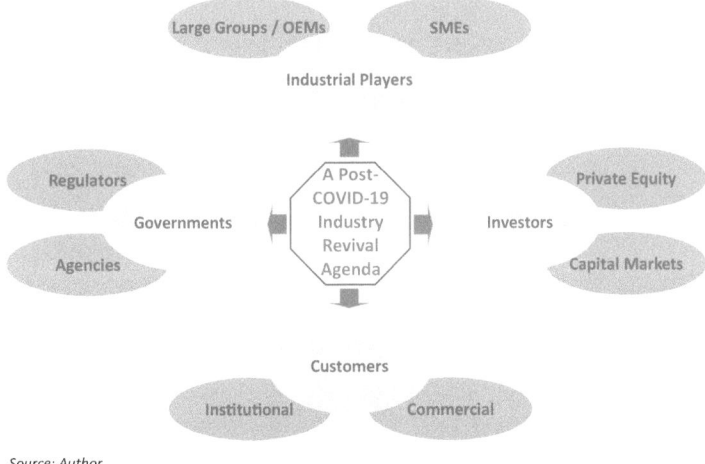

Source: Author

A Post-COVID-19 Agenda for Industry Revival

Here is a list of key themes and questions to be used as guidelines to find a new way forward and rethink the business of aerospace in the post-COVID-19 world, for each type of stakeholder. The most relevant column(s) of the book is (are) indicated in italic, next to each theme or question.

For all industrial players, large and small:

1. **Look at your environment from outside in**
   - What is happening around you (not just one degree of separation but two or three degrees of separation away)? *(re. 'Digital Aerospace')*
   - What major changes can you identify or anticipate? Can you sort out the signals from the noise? *(re. 'Game Over')*
   - Who are your real competitors? Is your market what you thought it was? *(re. 'Digital Aerospace', 'Game Over')*
   - Do you see some potential disruptors coming in, even in remote markets? *(re. 'Disruptive Innovations')*

2. **Revisit your vision and purpose**
   - A vision must stretch the imagination, must aim high and far, and possibly challenge conventional wisdom *(re. 'Musk-Disney', 'Creative Destruction', 'Customer-Supplier Relationship')*
   - A purpose should be close to people's hearts, something easy to connect with at individual level *(re. 'The Age of Resilience', 'The Purpose of Space Exploration', 'The Lost Art of Air Travel')*

3. **Refocus on core values and what makes your company unique**
   - Do not underestimate the importance of strong local roots is shaping your company's future *(re. 'Clusters')*
   - Refocus on the core business and protect your market shares before venturing into new markets *(re. 'Bombardier', 'Game Over')*
   - Make sure your value proposition is clear inside and outside the company *(re. 'L3', 'Creative Destruction', 'Comfort Zone')*

4. **Go back to the fundamentals (of business and strategy)**
   - A strong corporate culture goes a long way…especially when times are hard *(re. 'Culture is Destiny', 'Rolls Royce', 'Airbus' Innovation Gamble', 'Age of Resilience')*

- Be authentic, do not just follow fads – do not be afraid of being an outlier *(re. 'Dassault')*
- <u>Craftsmanship</u> is key, even in a high-tech. data-driven industry *(re. 'Closing the Deal', 'Dassault')*
- Manage the political risk proactively (think 'co-opetition') *(re. 'Managing the Political Risk')*
- Remember that a well-crafted and well-executed strategy pays off *(re. 'L3')*
- Remember also that every organization goes through cycles similar to the four seasons: plant the seeds of future growth early enough not to be stuck in winter *(re. 'L3')*
- Do not forget to analyze your market and industry dynamics before making strategic decisions *(re. 'Bombardier')*

5. Use M&A for the right reasons
- Stay away from financial engineering but focus on industrial and commercial value creation from the customers' point of view *(re. 'Out of Steam', 'Harris–Exelis')*
- Use M&A as an opportunity for strategic reframing, for redefining the company's playing field or moving the business boundaries to stimulate innovation and develop synergies that ultimately benefit customers *(re. 'Harris–Exelis', 'Safran-Zodiac')*
- Ask the right questions: Will it create value, or will it just be a zero-sum game? Will it help customers "do their job" better? Is it driven by true stewardship and strategic vision? *(re. 'Safran–Zodiac', 'Harris-Exelis', 'Out of Steam')*

6. Take your destiny into your own hands
- Do not wait for your (institutional) customers to support your international sales *(re. 'Merkel's Mark')*
- Create your own wave(s) (do not just ride whatever wave everybody else is riding) *(re. 'Out of Steam')*
- Look at your environment and see whether there is an opportunity to create a market space for yourself, to carve out a unique niche that would allow you to capture a disproportionate share of the market *(re. 'L3')*
- Do not rely on your (institutional) customers to fund your R&D, embrace commercially driven innovation *(re. 'Disruptive Imperative')*

7. **Innovate wisely**
    - Create your own "ecosystem" if you have to *(re. 'Disruptive Innovations', 'Creative Destruction')*
    - Do not cram disruptive innovations in your mainstream organization or products *(re. 'Airbus' Innovation Gamble')*
    - Recognize potential areas for disruptive innovation, where performance dimensions valued by customers are changing, creating opportunities for new players to come in with different value propositions *(re. 'Disruptive Innovations', 'Comfort Zone')*
    - Approach your innovation agenda like Silicon Valley start-ups: be long on vision and talent and short on time and budget *(re. 'Disruptive Imperative')*
    - Be sensible and methodical in your creative process: Explore things from different perspectives *(re. 'Musk – Disney')*
    - Understand where you are on the technology maturity curve and innovation adoption cycle *(re. 'Nanosatellites')*
    - Think of innovation as a trade-on opportunity (e.g., to move to a new cost curve) *(re. 'Aerostructures')*

For OEMs, Large Groups:

8. **Review your business portfolio and manage it dynamically**
    - Get rid of dis-economies of scale, do not be afraid to divest or break up *(re. 'Shrinking to Grow', 'Airbus Shows the Way')*
    - Make sure the whole is worth more than the sum of the parts *(re. 'Safran–Zodiac')*
    - Make sure your portfolio is well balanced (look for counter-cyclicality, identify lame ducks / question marks) *(re. 'Airbus at 50', 'A and D Divergence', 'Shrinking to Grow')*
    - Keep in mind that bigger is not necessarily better *(re. 'The Big White Space')*; vertical integration is not the only way to create value *(re. 'L3', 'New Predators', 'Private Equity's Promise')*
    - Simplify, get rid of excessive bureaucracy *(re. 'Airbus at 50')*

9. **Revisit your supply chain strategy and engage in a new customer-supplier relationship model**
    - Recognize what makes your supply chain competitive *(re. 'Chain Drive', 'Customer-Supplier Relationship')*

- Involve your suppliers early on, build mutual trust and long-term commitment *(re. 'Customer-Supplier Relationship')*
- Get to know your suppliers better, "walk the talk", nurture "supplier intimacy" *(re. 'Chain Drive')*

For governments:

10. Implement a proper national industrial strategy
- Nurture your industrial and technological base from the ground up, starting with start-ups and small companies *(re. 'The Big White Space', 'The New Predators')*
- Do not assume that your industry only needs a few big players to be strong and resilient. Create the necessary incentives to help small companies become big enough to stand their own ground (vs. large groups) *(re. 'Private Equity's Promise', 'The Big White Space', 'Separating Hype from Reality')*
- Be pragmatic and forward-thinking about international cooperation and export policies *(re. 'Galileo's High Stakes', 'Merkel's Mark')*

For institutional customers:

11. Be 'smart' customers
- Be in tune with market needs, do not just think in terms of national ego *(re. 'Merkel's Mark', 'Separating Hype from Reality')*
- Engage with the private sector with a forward-looking, value-adding partnership model *(re. 'Mars')*
- Let new, more diverse players compete on a level playing field *(re. 'Comfort Zone')*
- Change the way you procure your products and services; do not be too conservative *(re. 'Comfort Zone')*

For commercial customers (air travelers):

12. Be more discerning customers
- Think twice before flying, whether you are the one paying or not *(re. 'Dirty Secret')*
- Be more demanding in terms of customer experience and service quality, even if you fly 'economy' – do not let air travel become a commodity *(re. 'The Lost Art of Air Travel')*
- Think twice before buying a 'low-fare' ticket and reflect on the true value (and cost) of flying *(re. 'Dirty Secret', 'The Age or Resilience')*

For (private equity) investors:

13. Change your investment model
- Be more patient (invest for 5-10 years rather than 3-5 years) and more ambitious (aim for 5-6x growth, not just 2-3x) *(re. 'Private Equity's Promise', 'The New Predators', 'Industry-Finance Gap')*

For all stakeholders:

14. Embrace the new age of aviation: The Age of Resilience
- Let us redefine air travel – bring back some humanity and creativity in the business, possibly moving towards a new paradigm *(re. 'The Lost Art of Air Travel', 'Smaller, Slower, Closer')*
- Let us change our individual behaviors to make air travel more sustainable *(re. 'Dirty Secret', 'The Age or Resilience')*
- Let us use our creativity to envision the aviation ecosystem of tomorrow in a way that transforms constraints into opportunities and makes it a believable "win-win" value proposition for both the aerospace community and the broader world's community *(re. 'Customer-Supplier Relationship')*

Of course, these items are mainly directional and aspirational, and as such need to stand the test of real-life practicality. Nevertheless, they do offer a coherent and holistic framework to capture and make sense of the forces at play and that is the essence of strategic thinking.

Strategy is first and foremost about forces, forces on which an organization depends and forces it can influence. Whatever their nature, these forces are the raw material with which corporate leaders, entrepreneurs, lawmakers, and investors can shape their view of the world and act accordingly.

In the post-COVID-19 world, more than ever, all stakeholders will need strategic vision, as well as practical wisdom and creativity, to make the business of aerospace as relevant to the rest of the 21st century as it has been to the 20th century. The next couple of years are going to be painful for all and fatal for some, but we can only hope that, over the next decade, most aerospace players will prove resourceful enough to look like phoenixes rising from the ashes rather than swans singing their last song. In doing so, they will not only revive the industry's fortunes but also the spirit of its founders, who, as per Lindbergh's words, were driven by *"the love of the air and sky and flying, the lure of adventure, the appreciation of beauty"*.

A POST-COVID-19 AGENDA FOR INDUSTRY REVIVAL

For all industrial players, large and small:

1. Look at your environment from outside in
2. Revisit your vision and purpose
3. Refocus on core values and what makes your company unique
4. Go back to the fundamentals (of business and strategy)
5. Use M&A for the right reasons
6. Take your destiny into your own hands
7. Innovate wisely

For large groups / OEMs:

8. Review your business portfolio and manage it dynamically
9. Revisit your supply chain strategy and engage in a new customer-supplier relationship model

For governments:

10. Implement a proper national industrial strategy

For institutional customers:

11. Be 'smart' customers

For commercial customers (air travelers):

12. Be more discerning customers

For private equity investors:

13. Change your investment model

For all stakeholders:

14. Embrace the new age of aviation: The Age of Resilience

INDEX

"Key Concepts" in italic

accretive effect, 15, *17*
ACE Management, 37
AECC. *See* Aero Engine Corp. of China
Aero Engine Corp. of China (AECC), 48–49
Aerojet Rocketdyne, 146
aerospace and defense (A&D) industry
   British, 54–56
   COVID-19 pandemic, 158, 173–179
   craftsmanship, 63–64, 71–73
   divergence, 18–20
   future of, 157–172
   geographic clusters, 63, 68–70
   geopolitics in, 47–59
   growth strategies, 79–92
   innovation in, 119–137
   political risk, 64, 74–77
   structure and competition, 3–31
      aerostructures, 24–26
      Airbus Group *vs.* Boeing, 27–31
      component and subsystem strategy, 5, 15–17
      digital economy, 21–23
      Europe, 3–5, 7–31
      US, 3–5, 7–31
   successes and failures, case studies on, 93–115
   supply chain, 33–45
aerostructures, 24–26
air travel, 159–163
Airbus Group, 18–19
   vs. Boeing, 27–31
   vs. digital natives, 21, 22
   vs. Lockheed Martin, 18, 19, 21
   Bombardier C-Series, 110–112
   business portfolio, 79, 84–86
   corporate story, 106–108

   divestments, 84–86
   in China, 48, 49
   in the UK, 55, 56
   innovation strategy, 119, 135–137
   successes and failures (case study), 94, 106–109
   supply chain (structural weaknesses), 37–39
aircraft engines, 48, 49, 91
   rhenium, 48
   turbine blades, 48–49
air travel, 157, 159–163
Apple, 21–22, 125, 126
Ariane (space launcher), 108, 141
aviation ecosystem, 124, 125, 165

BAE Systems, 11, 50, 55, 88
   British aerospace, 55–56
   corporate strategy, 93, 95–97
   corporate story, 95–97
   EADS tie-up, 11, 12
   successes and failures (case study), 93, 95–97
   transatlantic consolidation, 88
B/E Aerospace, 127, 128
Belyamani, Seddik, 71
Bezos, Jeff, 128, 146, 154
big data, 73
Blue Origin, 124, 146, 154
Boeing, 18–19
   737MAX, 27, 158, 170–171
   vs. Airbus (Group), 27–31
   vs. digital natives, 21, 22
   vs. Lockheed Martin, 18, 19
   JDAM, 120–121
   value propositions, 131
Bombardier, 94, 110–112

181

Brexit, 50, 54–57, 59
British aerospace industry, 54–56
business ecosystem, 126
business model, 148–149
business performance (profitability), 63, 65–67
business portfolio management, 79, 83
    Airbus, 84–86
    General Dynamics, 81–83
    Thales, 81–82

capitalism, 87, 90, 99
capital markets, 173
Carlyle Group, 7–8
Chabraja, Nicholas, 82
China, 29, 30, 48–49
    AECC, 48–49
    AVIC, 48
    COMAC, 29, 30, 48, 49, 112
    NDRC, 49
Christensen, Clay, 119–123
closed vs. open innovation, 134
Clusters (geographic), 68–70
cluster effect, 69–70
cluster(s), 63, 68–70
COMAC, China, 29–30, 49
commercial aerospace
    Airbus *vs.* Boeing, 27–31
    Boeing and Airbus, 27–31
    Bombardier, 92, 110–112
    China, 29–30, 48–49
    COVID-19 pandemic, 156, 173–179
    vs. defense sector, 18–19
    next development phase, 24
    stability, 127–129
commercial innovation model, 119, 124–126
commercial-off- the-shelf (COTS)
    technology, 133
community, 126
Components & Sub-Systems, 4, 33, 35, 104–105
COMAC, 29, 30, 48, 49, 112
co-opetition, 76–77
corporate culture, 63, 65–67, 68, 136, 174
    Airbus Group, 27–30, 85, 135, 136
    Boeing 737MAX, 27, 70, 158
    Rockwell Collins, 63, 65–66
    Rolls-Royce, 113–115
    Zodiac Aerospace, 63, 66–67
corporate strategy, 63–115
    value proposition(s), 105, 128, 130–131, 174, 176, 178

corporate venture capital (CVC), 137
corporate venturing, 137
cost curve(s), 26
counter-cyclicality, 20
COVID-19 pandemic, 27, 30, 158, 173–179
    post COVID-19 agenda, 174–179
    Post COVID-19 agenda for industry revival, 173–179
craftsmanship, 63, 71–73, 93, 99, 100
C Series program, 110–112
CubeSats, 121–123, 142, 150–152
customer behavior, 127–129
    closed systems, 127–128
    conservatism, 99, 100, 128, 129
customer experience, 159–160
customer-supplier relationship model, 43–45
CVC. *See* Corporate venture capital

Dassault Aviation, 11–12, 93, 98–100
    Mirage 2000, 53
Dassault's Mirage 2000 aircraft, 53
defense market, 11, 50, 72, 88–89, 98
digital economy, 21–23
Disney, Walt, 139, 144–145
disruptive innovation model, 119, 120–123
divestments, 84–85
Drake's equation, 168
DRS Technologies, 7
duopoly, 31
dynamic business portfolio management, 79, 83
    Airbus, 79, 84–86
    General Dynamics, 79, 81–83
    Thales, 79, 81–82

EADS, 11
EADS-BAE tie-up, 11–12
economic touchpoints, 22–23
economies vs. dis-economies of scale, 14
ecosystem, business, 126
Enders, Tom, 85, 135–136
Europe A&D industry
    "big white space", 5
    private equity investors, 5, 7–10
    industry structure, 3–5
        component and subsystem strategy, 15–17
        divergence, 18–20
        need for change, 3–5
        sovereignty, 5, 17
European Union (EU), 47, 57–59, 115

INDEX

Exelis, 80, 87–89
export
  deals, 52
  market, 52–53
  policy, 47, 52
  regulations (US ITAR), 74–75

Figeac Aero, 38
fighter aircraft, 19, 52, 54, 99, 132
financial engineering, 35, 87
financial gap, 35–36
financial markets, 36
Finmeccanica, 11
Flir Systems, 133
France, 74–75
France, Rafale aircraft, 53
future of aviation, 157–172
  air travel, 157, 159–163
  post-COVID-19 agenda for industry revival, 173–179
  post COVID-19 pandemic, 158, 173–179
  space exploration, 158, 167–168
  urban air mobility, 157, 164–166
  zero-emission goal, 157, 161–162

Galileo satellite program, 57–58
GE Aviation, 18–19, 69, 73
Genchi Genbutsu, 41, *42*
General Dynamics (GD), 79, 81–83
geopolitics in A&D industry, 47–59
  British, 54–56
  China, 48–49
  France, 53
  Galileo program, UK, 57–58
  German, 52–53
  UK, 50–51, 54–56, 57–58
Germany ('Merkel doctrine'), 52–53
Gerondeau, Jean-Louis, 66
GKN, 51, 55–56
Global Navigation Satellite System (GNSS), 57–58
GNSS. *See* Global Navigation Satellite System
GOMspace, 150–151
Google, 21–23, 125, 135, 142, 150
greenhouse gas (GHG) emissions, 161
growth strategies, A&D industry, 79–92
  dynamic business portfolio management, 79, 81–86
  Mergers &Acquisitions (M&A), 79–80, 87–92

Harris Corporation, 80, 87–89
hedge funds, 35, 90
Hives, Ernest, 114
Hollande, Francois, 75
horizontal *vs.* vertical integration, 104–105
Hsieh, Tony, 67
hypersonic travel, 159, 160

Indra Sistemas, 8
industry cost curve, 26
industry-finance gap, 35–36
industry maturity, 151
innovation cycle, 152–153
innovation strategy, A&D industry, 119–137
innovation (models), 119–137
  Airbus Group, 135–137
  closed vs. open innovation, 134
  commercial innovation model, 119, 124–126
  disruptive, 119, 120–123
  obstacles to, 119, 127–134
integration vs. dis-integration cycles, 10
International Traffic in Arms Regulations (ITAR), 74–75
ITAR. *See* International Traffic in Arms Regulations

JDAM. *See* Joint Direct Attack Munition
Joint Direct Attack Munition (JDAM), 120–121
Jones, Clay, 65

keiretsu, 41, 44

L-3 Communications/L3 Technologies, 7, 12, 93, 101–105, 128
La Penta, Robert, 101
Lanza, Frank, 93, 101
Latécoère, 38
Leahy, John, 72
Levy, Jean-Bernard, 81
Lindbergh, Charles, 164, 166
Lockheed Martin, 7, 15, 18–19, 21, 81, 82, 98, 127, 146
long-term performance, 44

market attractiveness (assessment), 112
Mars, 140, 154–155, 167
McDonnell Douglas, 27

Mergers & Acquisitions (M&A), 79–80, 87–92
  Harris Corporation with Exelis, 80, 87–89
  Safran-Zodiac merger, 80, 90–92
Mitchell, Edgar, 168, 169
Moon, 154, 167–169
Musk, Elon, 22, 121, 128, 135, 143–145, 154–155, 167–168

nanosatellite industry, 140, 150–153
NASA, 132, 141, 143, 154–155, 167–168
National Development and Reform Commission (NDRC), 49
NDRC. *See* National Development and Reform Commission
New Glenn, 154
New Space, 139–156
  to colonize Mars, 140, 154–155
  master disruptor, 139, 143–145
  Musk, 143–145
  nanosatellite industry, 140, 150–153
  vs. old space, 139, 146–149
  public *vs.* private sectors partnership, 140, 154–155
  SpaceX, 121, 125, 128, 132, 139, 141–143

Obama, Barack, 75
Odyssey Investment Partners, 7
OEMs. *See* Original Equipment Manufacturers, 25, 37–39, 41–42, 68–69, 176, 179
OneWeb satellites, 18, 108, 135, 136
Onex, 7
Original Equipment Manufacturers (OEMs), 25, 38–39
  geographic clusters, 63, 68–70
outside-in *vs.* inside-out perspective, 23

paradigm shift, 146
partnership model (UK-EU), 59
Peace Corps Act of 1961, 155
Penta, Robert La, 101
Perry, Robert, 99
physical environment, 126
Piccard, Bertrand, 165
Planet Labs (Planet), 150
political risk, 64, 74–77
PPP. *See* Public private partnership, 154
private equity, 35–36
private equity investors

industry restructuring, role in, 5, 7–10
  vs. public companies, 15–17, 35
product portfolio assessment, 109
public private partnership (PPP), 154
public *vs.* private sectors partnership, 140, 154–155
Putin, Vladimir, 74

R&D (financing), 132–134
Rafale (aircraft), 52–53, 74, 75, 98–100
Raytheon, 15, 16, 82, 103, 127
resilience, 20, 51, 67, 113, 170–171, 178
Rockwell Collins, 63, 65–66, 73
Rolls-Royce, 50, 51, 55, 69, 92, 113–115

S3 (Swiss Space Systems), 142
Saab, 72
Safran, 11–12, 15–16, 35–36, 55, 66, 82, 83, 90–92, 106, 115, 127
  and Thales, 11
  Zodiac merger, 80, 90–92
Sagem, 12, 35, 66, 91
sales and marketing, 71–73
Schwartz, Barry, 160
"see-believe-think-act" problem-solving model, 44
Silicon Valley, 133–134, 135–136, 150
Sky and Space Global, 150–151
Skybox Imaging, 132–133, 142, 150
space exploration, 158, 167–169
SpaceX, 121, 125, 128, 132, 139, 141–143
Spire Global, 150, 151
Spirit Aerosystems, 7
successes and failures (case studies), A&D companies, 93–115
  Airbus, 94, 106–109
  BAE Systems, 93, 95–97
  Bombardier, 94, 110–112
  Dassault Aviation, 93, 98–100
  L-3 Communications, 93, 101–105
  Rolls-Royce, 94, 113–115
supply chain competitiveness, 40–42
supply chain, A&D industry, 33–45
  competitiveness, misconceptions of, 40–42
  customer-supplier relationship model, 43–45
  drivers, 40–42
  financial markets and, 35–36
  structural weaknesses, 37–39

supply chain structure, 39
sustainability (of air travel), 161–163, 170–171

TCI, 35, 90–91
technology maturity curve, 152–153
technology curve/innovation cycle, 152–153
Teledyne Technologies, 15–17
Thales, 11, 12, 16, 50, 81–83, 127
"think-act" problem-solving model, 44
three-dimensional printing, 125
ThyssenKrupp Marine Systems (TKMS), 52
tier-one suppliers, 39
TKMS. *See* Thyssen Krupp Marine Systems
Toyota Production System, 42
TransDigm, 7, 15–17

UK (United Kingdom), 50–51, 54–56, 57–59
Ultra Electronics, 8, 13
United Launch Alliance (ULA), 146
urban air mobility, 157, 163–165
U.S. A&D industry, 3–5, 7–10, 11, 50, 72, 88–89, 98

private equity investors, 5, 7–10
structure, 3–5, 7–10
  consolidation, 7–10
  reconfiguration process, 4
UTC (United Technologies), 15, 16, 73, 115

value chain, A&D, 4, 8, 10, 23, 31, 49, 65, 81, 102, 104–105, 128, 130
value-creation events, 21–23
  economic touchpoints, 22–23
value network, 126
value proposition, 101, 125, 128, *130–131*, 141, 142, 148–149, 174, 176, 178
Veritas Capital, 7
vertical-takeoff-and-landing (VTOL) vehicles, 164–165
vertical vs. horizontal integration, 104–105
Vought Aircraft Industries, 7
VT Group, 96

zero-emission, 157, 161–163
Zodiac Aerospace, 36, 66, 90–92
  corporate culture, 63, 66–67
  Safran merger, 80, 90–92

Acknowledgements

I would like to thank all my clients for the opportunity they gave me to work with them. This book would obviously not exist without them as they provided me with the real-life materials necessary to perform proper analyses and draw meaningful strategic insights.

My thanks also go to my consulting colleagues and partners whom I worked with over the years, at Booz Allen Hamilton, Candesic and Paragon European Partners. Consulting always works better and is more fun as a team.

I am very grateful to the Aviation Week & Space Technology team who gave me complete freedom to express my views in their magazine. My thoughts go particularly to late journalist Pierre Sparaco – who introduced me to the magazine, Tony Velocci, former editor-in-chief, who invited me to be part of a council of advisors for Aviation Week's annual Top Performing Companies rankings, and Joe Anselmo, current editor-in-chief, who offered me to write a regular opinion column for the magazine.

Finally, I would like to thank my long-time business partner – Philippe Cothier – whose support and loyalty have made this journey much more than a professional one. Indeed, as you may infer from reading this book, I strongly believe that behind every professional achievement, managerial decision, or engineering feat, however technology- or data-driven it may be, the human factor remains the ultimate differentiator.